Intelligent Optimisation Techniques

Springer
London
Berlin
Heidelberg
New York
Barcelona
Budapest
Hong Kong
Milan
Paris
Santa Clara
Singapore
Tokyo

D.T. Pham and D. Karaboga

Intelligent Optimisation Techniques

Genetic Algorithms, Tabu Search, Simulated Annealing and Neural Networks

With 115 Figures

 Springer

D. T. Pham, BE, PhD, DEng
Systems Division, School of Engineering, University of Wales, Cardiff, PO Box 688, Newport Road, Cardiff, CF2 3TE, UK

D. Karaboga, BSc MSc, PhD
Department of Electronic Engineering, Faculty of Engineering, Erciyes University, Kayseri, Turkey

ISBN 1-85233-028-7 Springer-Verlag Berlin Heidelberg New York

British Library Cataloguing in Publication Data
Pham, D. T. (Duc Truong), 1952-
    Intelligent optimisation techniques : genetic algorithms,
    tabu search, simulated annealing and neural networks
    1.Neural networks (Computer science) 2.Genetic algorithms
    3.Simulated annealing (Mathematics)
    I.Title II.Karaboga, Dervis
    006.3'2
    ISBN 1852330287

Library of Congress Cataloging-in-Publication Data
Pham, D.T.
        Intelligent optimisation techniques : genetic algorithms, tabu
    search, simulated annealing and neural networks / Duc Truong Pham
    and Dervis Karaboga
            p.   cm.
    Includes bibliographical references.
    ISBN 1-85233-028-7 (alk. paper)
        1. Engineering--Data processing. 2. Computer-aided engineering.
    3. Heuristic programming. 4. Genetic algorithms. 5. Simulated
    annealing (mathematics) 6. Neural networks (Computer science)
    I. Karaboga, Dervis, 1965- . II. Title.
    TA345.P54 1998
    620'.00285--dc21   98-15571

Typesetting: Camera ready by authors
Printed and bound at the Athenæum Press Ltd., Gateshead, Tyne & Wear
69/3830-543210 Printed on acid-free paper SPIN 10657875

Preface

This book covers four optimisation techniques loosely classified as "intelligent": genetic algorithms, tabu search, simulated annealing and neural networks.

- Genetic algorithms (GAs) locate optima using processes similar to those in natural selection and genetics.
- Tabu search is a heuristic procedure that employs dynamically generated constraints or tabus to guide the search for optimum solutions.
- Simulated annealing finds optima in a way analogous to the reaching of minimum energy configurations in metal annealing.
- Neural networks are computational models of the brain. Certain types of neural networks can be used for optimisation by exploiting their inherent ability to evolve in the direction of the negative gradient of an energy function and to reach a stable minimum of that function.

Aimed at engineers, the book gives a concise introduction to the four techniques and presents a range of applications drawn from electrical, electronic, manufacturing, mechanical and systems engineering. The book contains listings of C programs implementing the main techniques described to assist readers wishing to experiment with them.

The book does not assume a previous background in intelligent optimisation techniques. For readers unfamiliar with those techniques, Chapter 1 outlines the key concepts underpinning them. To provide a common framework for comparing the different techniques, the chapter describes their performances on simple benchmark numerical and combinatorial problems. More complex engineering applications are covered in the remaining four chapters of the book.

Chapter 2 comprises two sections. The first section presents four variations to the standard GA. The second section describes different GA applications, namely, design of fuzzy logic controllers, gearboxes and workplace layouts and training of recurrent neural networks for dynamic system modelling.

Chapter 3 studies the use of tabu search for designing microstrip antennas, training recurrent neural networks, designing digital FIR filters and tuning PID controller parameters.

Chapter 4 describes an application of simulated annealing to a real-time optimisation problem in the manufacture of optical fibre couplings. The chapter also reports on two other manufacturing engineering applications of simulated annealing, one concerned with the allocation of inspection stations in a multi-stage production system and the other with the selection of optimum lot sizes for batch production.

Chapter 5 outlines the use of neural networks to the problems of VLSI component placement and satellite broadcast scheduling.

In addition to the main chapters, the book also has six appendices. Appendices A1 and A2, respectively, provide background material on classical optimisation techniques and fuzzy logic theory. Appendices A3 to A6 contain the listings of C programs implementing the intelligent techniques covered in the book.

The book represents the culmination of research efforts by the authors and their teams over the past ten years. Financial support for different parts of this work was provided by several organisations to whom the authors are indebted. These include the Welsh Development Agency, the Higher Education Funding Council for Wales, the Engineering and Physical Science Research Council, the British Council, the Royal Society, the European Regional Development Fund, the European Commission, Federal Mogul (UK), Hewlett-Packard (UK), IBS (UK), Mitutoyo (UK), SAP (UK), Siemens (UK) and the Scientific and Technical Research Council of Turkey. The authors would also like to thank present and former members of their laboratories, in particular, Professor G. G. Jin, Dr A. Hassan, Dr H. H. Onder, Dr Y. Yang, Dr P. H. Channon, Dr D. Li, Dr I. Nicholas, Mr M. Castellani, Mr M. Barrere, Dr N. Karaboga, Dr A. Kalinli, Mr A. Kaplan, Mr R. Demirci and Mr U. Dagdelen, for contributing to the results reported in this book and/or checking drafts of sections of the book. The greatest help with completing the preparation of the book was provided by Dr B. J. Peat and Dr R. J. Alcock. They unstintingly devoted many weeks to editing the book for technical and typographical errors and justly deserve the authors' heartiest thanks.

Finally, the authors extend their appreciation to Mr A. R. Rowlands of the Cardiff School of Engineering for proofreading the manuscript and to Mr N. Pinfield and Mrs A. Jackson of Springer-Verlag London for their help with the production of the book and their patience during its long period of incubation.

D. T. Pham
D. Karaboga

Contents

Chapter 1

Introduction

This chapter consists of six main sections. The first four sections briefly introduce the basic principles of genetic algorithms, tabu search, simulated annealing and neural networks. To give an indication of the relative performances of these techniques, the last two sections present the results obtained using them to optimise a set of numeric test functions and a travelling salesman problem.

1.1 Genetic Algorithms

1.1.1 Background

Conventional search techniques, such as hill-climbing, are often incapable of optimising non-linear multimodal functions. In such cases, a random search method might be required. However, undirected search techniques are extremely inefficient for large domains. A genetic algorithm (GA) is a directed random search technique, invented by Holland [Holland, 1975], which can find the global optimal solution in complex multi-dimensional search spaces. A GA is modelled on natural evolution in that the operators it employs are inspired by the natural evolution process. These operators, known as genetic operators, manipulate individuals in a population over several generations to improve their fitness gradually. As discussed in the next section, individuals in a population are likened to chromosomes and usually represented as strings of binary numbers.

The evolution of a population of individuals is governed by the "schema theorem" [Holland, 1975]. A schema represents a set of individuals, i.e. a subset of the population, in terms of the similarity of bits at certain positions of those individuals. For example, the schema 1*0* describes the set of individuals whose first and third bits are 1 and 0, respectively. Here, the symbol * means that any value would be acceptable. In other words, the values of bits at positions marked * could be either 0 or 1 in a binary string. A schema is characterised by two parameters: defining length and order. The defining length is the length between the first and last bits with fixed values. The order of a schema is the number of bits with specified values. According to the schema theorem, the distribution of a schema through the population from one generation to the next depends on its order, defining length and fitness.

GAs do not use much knowledge about the problem to be optimised and do not deal directly with the parameters of the problem. They work with codes which represent the parameters. Thus, the first issue in a GA application is how to code the problem under study, i.e. how to represent the problem parameters. GAs operate with a population of possible solutions, not only one possible solution, and the second issue is therefore how to create the initial population of possible solutions. The third issue in a GA application is how to select or devise a suitable set of genetic operators. Finally, as with other search algorithms, GAs have to know the quality of already found solutions to improve them further. Therefore, there is a need for an interface between the problem environment and the GA itself for the GA to have this knowledge. The design of this interface can be regarded as the fourth issue.

1.1.2 Representation

The parameters to be optimised are usually represented in a string form since genetic operators are suitable for this type of representation. The method of representation has a major impact on the performance of the GA. Different representation schemes might cause different performances in terms of accuracy and computation time.

There are two common representation methods for numerical optimisation problems [Michalewicz, 1992; Davis, 1991]. The preferred method is the binary string representation method. The reason for this method being popular is that the binary alphabet offers the maximum number of schemata per bit compared to other coding techniques. Various binary coding schemes can be found in the literature, for example, Uniform coding and Gray scale coding. The second representation method is to use a vector of integers or real numbers, with each integer or real number representing a single parameter.

When a binary representation scheme is employed, an important issue is to decide the number of bits used to encode the parameters to be optimised. Each parameter should be encoded with the optimal number of bits covering all possible solutions in the solution space. When too few or too many bits are used the performance can be adversely affected.

1.1.3 Creation of Initial Population

At the start of optimisation, a GA requires a group of initial solutions. There are two ways of forming this initial population. The first consists of using randomly produced solutions created by a random number generator. This method is preferred for problems about which no a priori knowledge exists or for assessing the performance of an algorithm.

The second method employs a priori knowledge about the given optimisation problem. Using this knowledge, a set of requirements is obtained and solutions which satisfy those requirements are collected to form an initial population. In this case, the GA starts the optimisation with a set of approximately known solutions and therefore converges to an optimal solution in less time than with the previous method.

1.1.4 Genetic Operators

The flowchart of a simple GA is given in Figure 1.1. There are three common genetic operators: selection, crossover and mutation. An additional reproduction operator, inversion, is sometimes also applied. Some of these operators were inspired by nature and, in the literature, many versions of these operators can be found. It is not necessary to employ all of these operators in a GA because each functions independently of the others. The choice or design of operators depends on the problem and the representation scheme employed. For instance, operators designed for binary strings cannot be directly used on strings coded with integers or real numbers.

Selection. The aim of the selection procedure is to reproduce more copies of individuals whose fitness values are higher than those whose fitness values are low. The selection procedure has a significant influence on driving the search towards a promising area and finding good solutions in a short time. However, the diversity of the population must be maintained to avoid premature convergence and to reach the global optimal solution. In GAs there are mainly two selection procedures: proportional selection and ranking-based selection [Whitely, 1989].

Proportional selection is usually called "roulette wheel" selection since its mechanism is reminiscent of the operation of a roulette wheel. Fitness values of individuals represent the widths of slots on the wheel. After a random spinning of the wheel to select an individual for the next generation, individuals in slots with large widths representing high fitness values will have a higher chance to be selected.

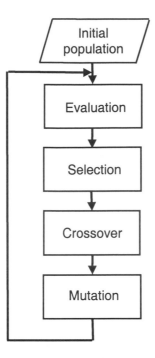

Fig. 1.1 Flowchart of a simple genetic algorithm

One way to prevent rapid convergence is to limit the range of trials allocated to any single individual, so that no individual generates too many offspring. The ranking-based production procedure is based on this idea. According to this procedure, each individual generates an expected number of offspring which is based on the rank of its fitness value and not on the magnitude [Baker, 1985].

Crossover. This operation is considered the one that makes the GA different from other algorithms, such as dynamic programming. It is used to create two new individuals (children) from two existing individuals (parents) picked from the current population by the selection operation. There are several ways of doing this. Some common crossover operations are one-point crossover, two-point crossover, cycle crossover and uniform crossover.

One-point crossover is the simplest crossover operation. Two individuals are randomly selected as parents from the pool of individuals formed by the selection procedure and cut at a randomly chosen point. The tails, which are the parts after the cutting point, are swapped and two new individuals (children) are produced.

Note that this operation does not change the values of bits. Examples are shown in Figure 1.2(a-d) for different crossover operations.

Mutation. In this procedure, all individuals in the population are checked bit by bit and the bit values are randomly reversed according to a specified rate. Unlike crossover, this is a monadic operation. That is, a child string is produced from a single parent string. The mutation operator forces the algorithm to search new areas. Eventually, it helps the GA avoid premature convergence and find the global optimal solution. An example is given in Figure 1.3.

Like other operators, many versions of this operator exist in the literature. In Chapter 2, some new mutation strategies are described in detail.

Parent 1	**1 0 0 0 1 0 0 1 1 1 1**
Parent 2	0 1 1 0 1 1 0 0 0 1 1
New string 1	**1 0 0 0 1** 1 0 0 0 1 1
New string 2	0 1 1 0 1 **0 0 1 1 1 1**

Fig. 1.2 (a) One-point crossover

Parent 1	**1 0 1 0 0 0 1 1 0 1 0**
Parent 2	0 1 1 0 1 1 1 1 0 1 1
New string 1	**1 0 1 0** 1 1 1 **1 0 1 0**
New string 2	0 1 1 **0 0 0 1 1** 0 1 1

Fig. 1.2 (b) Two-point crossover

Parent 1   1 2 3 4 5 6 7 8

Parent 2   **a b c d e f g h**

New string 1   1 **b** 3 **d e** 6 **g** 8

New string 2   **a** 2 **c** 4 5 **f** 7 **h**

Fig. 1.2 (c) Cycle crossover

Parent 1   1 0 0 1 0 1 1

Parent 2   **0 1 0 1 1 0 1**

Template   1 1 0 1 0 0 1

New string 1   1 0 **0** 1 **1 0** 1

New string 2   **0 1** 0 **1** 0 1 1

Fig. 1.2 (d) Uniform crossover

Old string   1 1 0 0 **0** 1 0 1 1 1 0

New string   1 1 0 0 **1** 1 0 1 1 1 0

Fig. 1.3 Mutation

Inversion. This additional operator is employed for a number of problems described in this book, including the cell placement problem, layout problems and the travelling salesman problem. It also operates on one individual at a time. Two points are randomly selected from an individual and the part of the string between those two points is reversed (see Figure 1.4).

Old string	1 0 **1 1 0 0** 1 1 1 0
New string	1 0 **0 0 1 1** 1 1 1 0

Fig. 1.4 Inversion of a binary string segment

1.1.5 Control Parameters

Important control parameters of a simple GA include the population size (number of individuals in the population), crossover rate and mutation rate. Several researchers have studied the effect of these parameters on the performance of a GA [Schaffer *et al.*, 1989; Grefenstette, 1986; Fogarty, 1989]. The main conclusions are as follows. A large population size means the simultaneous handling of many solutions and increases the computation time per iteration; however since many samples from the search space are used, the probability of convergence to a global optimal solution is higher than when using a small population size. The crossover rate determines the frequency of the crossover operation. It is useful at the start of optimisation to discover a promising region. A low crossover frequency decreases the speed of convergence to such an area. If the frequency is too high, it leads to saturation around one solution. The mutation operation is controlled by the mutation rate. A high mutation rate introduces high diversity in the population and might cause instability. On the other hand, it is usually very difficult for a GA to find a global optimal solution with too low a mutation rate.

1.1.6 Fitness Evaluation Function

The fitness evaluation unit acts as an interface between the GA and the optimisation problem. The GA assesses solutions for their quality according to the information produced by this unit and not by using direct information about their structure. In engineering design problems, functional requirements are specified to the designer who has to produce a structure which performs the desired functions within predetermined constraints. The quality of a proposed solution is usually calculated depending on how well the solution performs the desired functions and satisfies the given constraints. In the case of a GA, this calculation must be automatic and the problem is how to devise a procedure which computes the quality of solutions.

Fitness evaluation functions might be complex or simple depending on the optimisation problem at hand. Where a mathematical equation cannot be formulated for this task, a rule-based procedure can be constructed for use as a fitness function or in some cases both can be combined. Where some constraints are very important and cannot be violated, the structures or solutions which do so can be eliminated in advance by appropriately designing the representation scheme. Alternatively, they can be given low probabilities by using special penalty functions.

1.2 Tabu Search

1.2.1 Background

The tabu search algorithm was developed independently by Glover [1986] and Hansen [1986] for solving combinatorial optimisation problems. It is a kind of iterative search and is characterised by the use of a flexible memory. It is able to eliminate local minima and to search areas beyond a local minimum. Therefore, it has the ability to find the global minimum of a multimodal search space. The process with which tabu search overcomes the local optimality problem is based on an evaluation function that chooses the highest evaluation solution at each iteration. This means moving to the best admissible solution in the neighbourhood of the current solution in terms of the objective value and tabu restrictions. The evaluation function selects the move that produces the most improvement or the least deterioration in the objective function. A tabu list is employed to store the characteristics of accepted moves so that these characteristics can be used to classify certain moves as tabu (i.e. to be avoided) in later iterations. In other words, the tabu list determines which solutions may be reached by a move from the current solution. Since moves not leading to improvements are accepted in tabu search, it is possible to return to already visited solutions. This might cause a cycling problem to arise. The tabu list is used to overcome this problem. A strategy called the forbidding strategy is employed to control and update the tabu list. By using the forbidding strategy, a path previously visited is avoided and new regions of the search space are explored.

1.2.2 Strategies

A simple tabu search algorithm consists of three main strategies: forbidding strategy, freeing strategy and short-term strategy [Glover, 1989; Glover, 1990].

The forbidding strategy controls what enters the tabu list. The freeing strategy controls what exits the tabu list and when. The short-term strategy manages the interplay between the forbidding and freeing strategies to select trial solutions. Apart from these strategies, there can be also a learning strategy which consists in the use of intermediate and long-term memory functions. This strategy collects information during a tabu search run and this information is used to direct the search in subsequent runs.

Forbidding Strategy. This strategy is employed to avoid cycling problems by forbidding certain moves or in other words classifying them as tabu. In order to prevent the cycling problem, it is sufficient to check if a previously visited solution is revisited. Ideally, the tabu list must store all previously visited solutions and before any new move is carried out the list must be checked. However, this requires too much memory and computational effort. A simple rule to avoid the cycling problem could be not visiting the solution visited at the last iteration. However, it is clear that this precaution does not guarantee that cycling will not occur. An alternative way might be not visiting the solutions already visited during the last T_s iterations (these solutions are stored in the tabu list). Thus, by preventing the choice of moves that represent the reversal of any decision taken during a sequence of the last T_s iterations, the search moves progressively away from all solutions of the previous T_s iterations. Here, T_s is normally called the tabu list length or tabu list size. With the help of an appropriate value of T_s, the likelihood of cycling effectively vanishes. If this value is too small, the probability of cycling is high. If it is too large then the search might be driven away from good solution regions before these regions are completely explored.

The tabu list embodies one of the primary short-term memory functions of tabu search. As explained above, it is implemented by registering only the T_s most recent moves. Once the list is full each new move is written over the oldest move in the list. Effectively the tabu list is processed as a circular array in a first-in-first-out (FIFO) procedure.

Aspiration Criteria and Tabu Restrictions. An aspiration criterion is used to make a tabu solution free if this solution is of sufficient quality and can prevent cycling. While an aspiration criterion has a role in guiding the search process, tabu restrictions have a role in constraining the search space. A solution is acceptable if the tabu restrictions are satisfied. However, a tabu solution is also assumed acceptable if an aspiration criterion applies regardless of the tabu status. The move attributes are recorded and used in tabu search to impose constraints that prevent moves from being chosen that would reverse the changes represented by these attributes. Tabu restrictions are also used to avoid repetitions rather than reversals. These have the role of preventing the repetition of a search path that leads away from a given solution. By contrast, restrictions that prevent reversals have the role of preventing a return to a previous solution. A tabu restriction is typically activated only in the case where its attributes occurred within a limited number of iterations prior to the present iteration (a recency-based restriction), or occurred with a certain frequency over a larger number of iterations (a frequency-based restriction). More precisely, a tabu restriction is enforced only when the attributes underlying its definition satisfy certain thresholds of recency or frequency.

In recency-based restriction, a tabu duration is determined and the tabu solution is retained as tabu throughout the tabu duration. Rules for determining the tabu duration are classified as static or dynamic. Static rules choose a value for the duration that remains fixed throughout the search. Dynamic rules allow the value of the tenure to vary.

In frequency-based restriction, a frequency measure is used. The measure is a ratio whose numerator represents the count of the number of occurrences of a particular event and whose denominator generally represents one of the following quantities [Reeves, 1995]:

(a)   Sum of the numerators
(b)   Maximum numerator value
(c)   Average numerator value

The appropriate use of aspiration criteria can be very significant for enabling a tabu search to achieve its best performance. An aspiration criterion can be either time-independent or time-dependent. Early applications of tabu search employed only a simple type of aspiration criterion which is a time-independent criterion. It consists of removing a tabu classification from a trial solution when the solution shows better performance than the best obtained so far. This remains widely used. Another widely used aspiration criterion is aspiration by default. With this criterion, if all available moves are classified as tabu, and are not rendered admissible by other aspiration criteria, then the "least tabu" solution is selected. This could be a solution that loses its tabu classification by the least increase in the value of the present iteration number. Apart from these criteria, there are several other criteria used for aspiration such as aspiration by objective, aspiration by search direction and aspiration by influence [Reeves, 1995].

Freeing Strategy. The freeing strategy is used to decide what exits the tabu list. The strategy deletes the tabu restrictions of the solutions so that they can be reconsidered in further steps of the search. The attributes of a tabu solution remain on the tabu list for a duration of T_s iterations. A solution is considered admissible if its attributes are not tabu or if it has passed the aspiration criterion test.

Intermediate and Long-Term Learning Strategies. These strategies are implemented using intermediate and long-term memory functions. The intermediate function provides an element of intensification. It operates by recording good features of a selected number of moves generated during the execution of the algorithm. This can be considered a learning strategy which seeks new solutions that exhibit similar features to those previously recorded. This is achieved by restricting moves that do not possess favourable features.

Short-Term Strategy (Overall Strategy). This strategy manages the interplay between the above different strategies. The overall strategy is shown in Figure 1.5. A candidate list is a sub list of the possible moves. Candidate list strategies are generally problem dependent and can be derived by various methods such as random sampling.

The best-solution selection strategy selects an admissible solution from the current solutions if it yields the greatest improvement or the least deterioration in the objective function subject to the tabu restrictions and aspiration criterion being satisfied. This is an aggressive criterion and is based on the assumption that solutions with higher evaluations have a higher probability of either leading to a near optimal solution, or leading to a good solution in a fewer number of steps. If a solution satisfies the aspiration criterion it is admissible whether or not it is tabu. If a solution does not satisfy these criteria then it is admissible if it is not tabu.

A stopping criterion terminates the tabu search procedure after a specified number of iterations have been performed either in total, or since the current best solution was found.

1.3 Simulated Annealing

1.3.1 Background

The simulated annealing algorithm was derived from statistical mechanics. Kirkpatrick *et al.* [1983] proposed an algorithm which is based on the analogy between the annealing of solids and the problem of solving combinatorial optimisation problems.

Annealing is the physical process of heating up a solid and then cooling it down slowly until it crystallises. The atoms in the material have high energies at high temperatures and have more freedom to arrange themselves. As the temperature is reduced, the atomic energies decrease. A crystal with regular structure is obtained at the state where the system has minimum energy. If the cooling is carried out very quickly, which is known as rapid quenching, widespread irregularities and defects are seen in the crystal structure. The system does not reach the minimum energy state and ends in a polycrystalline state which has a higher energy.

At a given temperature, the probability distribution of system energies is determined by the Boltzmann probability:

$$P(E) \propto e^{[-E / (kT)]}$$ (1.1)

where E is system energy, k is Boltzmann's constant, T is the temperature and $P(E)$ is the probability that the system is in a state with energy E.

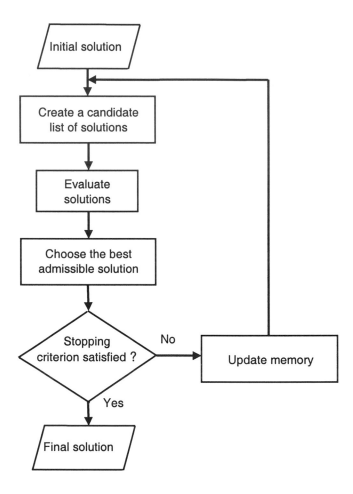

Fig. 1.5 Flowchart of a standard tabu search algorithm

At high temperatures, $P(E)$ converges to 1 for all energy states according to equation (1.1). It can also be seen that there exists a small probability that the system might have high energy even at low temperatures. Therefore, the statistical distribution of energies allows the system to escape from a local energy minimum.

1.3.2 Basic Elements

In the analogy between a combinatorial optimisation problem and the annealing process, the states of the solid represent feasible solutions of the optimisation problem, the energies of the states correspond to the values of the objective function computed at those solutions, the minimum energy state corresponds to the optimal solution to the problem and rapid quenching can be viewed as local optimisation.

The algorithm consists of a sequence of iterations. Each iteration consists of randomly changing the current solution to create a new solution in the neighbourhood of the current solution. The neighbourhood is defined by the choice of the generation mechanism. Once a new solution is created the corresponding change in the cost function is computed to decide whether the newly produced solution can be accepted as the current solution. If the change in the cost function is negative the newly produced solution is directly taken as the current solution. Otherwise, it is accepted according to Metropolis's criterion [Metropolis *et al.*, 1953] based on Boltzman's probability.

According to Metropolis's criterion, if the difference between the cost function values of the current and the newly produced solutions is equal to or larger than zero, a random number δ in $[0,1]$ is generated from a uniform distribution and if

$$\delta \leq e^{(-\Delta E/T)} \tag{1.2}$$

then the newly produced solution is accepted as the current solution. If not, the current solution is unchanged. In equation (1.2), ΔE is the difference between the cost function values of the two solutions.

The flowchart of a standard simulated annealing algorithm is presented in Figure 1.6. In order to implement the algorithm for a problem, there are four principal choices that must be made. These are:

(a) Representation of solutions
(b) Definition of the cost function
(c) Definition of the generation mechanism for the neighbours
(d) Designing a cooling schedule

Solution representation and cost function definitions are as for GAs. Various generation mechanisms could be developed that again could be borrowed from GAs, for example, mutation and inversion.

In designing the cooling schedule for a simulated annealing algorithm, four parameters must be specified. These are: an initial temperature, a temperature update rule, the number of iterations to be performed at each temperature step and a stopping criterion for the search.

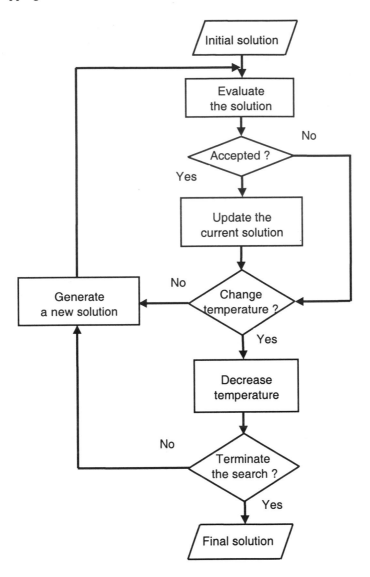

Fig. 1.6 Flowchart of a standard simulated annealing algorithm

There are several cooling schedules in the literature [Osman, 1991]. These employ different temperature updating schemes. Of these, stepwise, continuous and non-monotonic temperature reduction schemes are widely used. Stepwise reduction schemes include very simple cooling strategies. One example is the geometric cooling rule. This rule updates the temperature by the following formula:

$$T_{i+1} = c\, T_i\,, \quad i = 0,1...$$ (1.3)

where c is a temperature factor which is a constant smaller than 1 but close to 1.

1.4 Neural Networks

Neural networks are modelled on the mechanism of the brain [Kohonen, 1989; Hecht-Nielsen, 1990]. Theoretically, they have a parallel distributed information processing structure. Two of the major features of neural networks are their ability to learn from examples and their tolerance to noise and damage to their components.

1.4.1 Basic Unit

A neural network consists of a number of simple processing elements, also called nodes, units, short-term memory elements and neurons. These elements are modelled on the biological neuron and perform local information processing operations. A simple and general representation of a processing element is given in Figure 1.7. A processing element has one output and several inputs which could be its own output, the output of other processing elements or input signals from external devices. Processing elements are connected to one another via links with weights which represent the strengths of the connections. The weight of a link determines the effect of the output of a neuron on another neuron. It can be considered part of the long-term memory in a neural network.

After the inputs are received by a neuron, a pre-processing operation is applied. There are several alternative pre-processing procedures including taking the summation, cumulative summation, maximum or product of the weighted inputs. Some common functions are given in Table 1.1. The output of the pre-processing operation is passed through a function called the activation function to produce the final output of the processing element. Depending on the problem, various types of activation functions are employed, such as a linear function, step function, sigmoid function, hyperbolic-tangent function, etc. Some common activation functions are shown in Table 1.2.

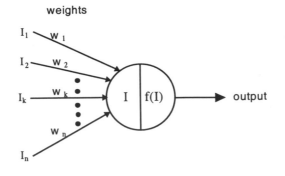

Fig. 1.7 Simple model of a neuron (Processing element)

Table 1.1 Some common pre-processing functions

SUMMATION	$I = \sum\limits_{k} w_k I_k$
PRODUCT	$I = \prod\limits_{k} w_k I_k$
CUMULATIVE SUMMATION	$I_{new} = I_{old} + \sum\limits_{k} w_k I_k$
MINIMUM	$I = \mathop{Min}\limits_{k}(w_k I_k)$
MAXIMUM	$I = \mathop{Max}\limits_{k}(w_k I_k)$

Table 1.2 Typical activation functions

LINEAR FUNCTION	$f(x) = x$
SIGMOID FUNCTION	$f(x) = 1 / (1 + e^{-x})$
THRESHOLD FUNCTION	$f(x) = \begin{cases} +1, & \text{if } x > x_t \\ -1, & \text{otherwise} \end{cases}$
HYPERBOLIC TANGENT FUNCTION	$f(x) = (e^x - e^{-x}) / (e^x + e^{-x})$
SINUSOIDAL FUNCTION	$f(x) = \text{Sin}(x)$

The general representation of an artificial neuron is a simplified model. In practice, neural networks employ a variety of neuron models that have more specific features than the general model. The Hopfield model is one of the most popular dynamic models for artificial neurons and is widely used in optimisation. The block diagram of a Hopfield neuron is shown in Figure 1.8. It contains a summer, an integrator and an activation function. The summer is a pre-processing component. It performs a summation operation on the weighted inputs $w_{ji} \cdot x_i$, with threshold θ_i. The activation function, $\psi(u_i)$, is normally a sigmoid function and is used to produce the output of the neuron. The damped integrator is located between the summer and the activation function. The damping coefficient α_i makes internal signal u_i zero for zero input of the neuron. This integrator provides the dynamic nature of the Hopfield neuron model whereby u_i depends not only on the current inputs but also on the previous ones.

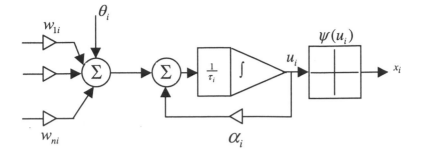

Fig. 1.8 Functional structure of a Hopfield neuron (Processing element)

How the inter-neuron connections are arranged and the nature of the connections determine the structure of a network. How the strengths of the connections are adjusted or trained to achieve a desired overall behaviour of the network is governed by its learning algorithm. Neural networks can be classified according to their structures and learning algorithms.

1.4.2 Structural Categorisation

In terms of their structures, neural networks can be divided into two types: feedforward networks and recurrent networks.

Feedforward Networks. In a feedforward network, the neurons are generally grouped into layers. Signals flow from the input layer through to the output layer via unidirectional connections, the neurons being connected from one layer to the next, but not within the same layer. Examples of feedforward networks include the multi-layer perceptron (MLP) [Rumelhart and McClelland, 1986], the learning vector quantisation (LVQ) network [Kohonen, 1989] and the group-method of data handling (GMDH) network [Hecht-Nielsen, 1990]. Feedforward networks can most naturally perform static mappings between an input space and an output space: the output at a given instant is a function only of the input at that instant. Feedforward networks normally employ simple static neurons.

Recurrent Networks. In a recurrent network the outputs of some neurons are fedback to the same neuron or to neurons in preceding layers. Thus, signals can flow in both forward and backward directions. Examples of recurrent networks include the Hopfield network [Hopfield, 1982], the Elman network [Elman, 1990] and the Jordan network [Jordan, 1986]. Recurrent networks have a dynamic memory: their outputs at a given instant reflect the current input as well as previous inputs and outputs. Elman and Jordan networks normally employ simple static neurons whereas Hopfield networks may use either simple static neurons or dynamic Hopfield neurons.

1.4.3 Learning Algorithm Categorisation

Neural networks are trained by two main types of learning algorithms: supervised and unsupervised learning algorithms. In addition, there exists a third type, reinforcement learning, which can be regarded as a special form of supervised learning.

Supervised Learning. A supervised learning algorithm adjusts the strengths or weights of the inter-neuron connections according to the difference between the desired and actual network outputs corresponding to a given input. Thus, supervised learning requires a "teacher" or "supervisor" to provide desired or target output signals. Examples of supervised learning algorithms include the delta rule [Widrow and Hoff, 1960], the generalised delta rule or backpropagation algorithm [Rumelhart and McClelland, 1986] and the LVQ algorithm [Kohonen, 1989]. Learning in the standard autoassociative Hopfield network may be considered a special case of supervised learning. The network employs a special one-step procedure during "learning" and an iterative procedure during recall. In contrast, most other networks utilise an iterative procedure during the learning phase.

Unsupervised Learning. Unsupervised learning algorithms do not require the desired outputs to be known. During training, only input patterns are presented to the neural network which automatically adapts the weights of its connections to cluster the input patterns into groups with similar features. Examples of unsupervised learning algorithms include the Kohonen [Kohonen, 1989] and Carpenter-Grossberg Adaptive Resonance Theory (ART) [Carpenter and Grossberg, 1988] competitive learning algorithms.

Reinforcement Learning. As mentioned before, reinforcement learning is a special case of supervised learning. Instead of using a teacher to give target outputs, a reinforcement learning algorithm employs a critic only to evaluate the goodness of the neural network output corresponding to a given input. An example of a reinforcement learning algorithm is the genetic algorithm (GA) [Holland, 1975; Goldberg, 1989].

1.4.4 Optimisation Algorithms

Optimisation problems can be solved by utilising neural networks that are designed inherently to minimise an energy function. This book in particular describes the use of the Hopfield network and the Kohonen self-organising network. In order to apply these networks to a particular optimisation problem, they must be tailored to the specific problem. This is done by coding the objective function (cost function) and constraints of the optimisation problem into the energy function of the network which aims to reach a stable state where its outputs yield the desired optimum parameters. This approach is applied to the optimisation problems at the end of this chapter and to those in Chapter 5.

Another approach, not detailed in this book, is to first train a neural model of the process to be optimised, using a suitable network such as an MLP, and then adapting the inputs to the model to produce the optimum process output. The input adaptation procedure can be the same as that for weight adaptation during the training of the neural model.

Rather than being seen as an optimisation tool, neural networks are more frequently the object *of* optimisation exercises. This is because the training of a neural network can be regarded as an optimisation problem. Such a problem could be solved by applying optimisation techniques such as GAs and tabu search. Work in this area is covered in Chapters 3 and 4, which report on the use of those techniques to train Elman and Jordan networks.

To reiterate, the neural network method used for solving optimisation problems focussed on in this book is for the neural networks to minimise a specified energy function. Because neural networks are often simulated and designed on the basis of system differential equations, gradient algorithms are suitable for implementing neural networks [Cichocki and Unbehauen, 1993].

Algorithms for unconstrained optimisation. Cost functions are energy functions in unconstrained optimisation. Gradient algorithms for this type of optimisation aim to satisfy a searching procedure along a gradient direction. In the continuous-time case, this direction can be considered as trajectory $\mathbf{x}(t)$ which can be determined from the general system of differential equations:

$$\frac{d\mathbf{x}}{dt} = -\mu(\mathbf{x},t) \cdot \nabla E(\mathbf{x}) \tag{1.4}$$

where $E(\mathbf{x})$ is the cost function and $\mu(\mathbf{x},t)$ is a symmetric positive definite matrix. There are different methods to determine the value of $\mu(\mathbf{x},t)$. For instance, it is calculated as $[\nabla^2 E(\mathbf{x})]^{-1}$ in Newton's algorithm, and it is chosen as $\mu(t)$ in the steepest-descent algorithm.

By using gradient algorithms, a search procedure will, in general, stop at a local minimum. To allow it to proceed continuously to a global minimum, the search procedure must be able to escape from local minima. There are several methods proposed for this purpose. The most commonly used method is to add a noise sequence onto the cost function. The gradient algorithm, for example, the steepest-descent algorithm, in this case becomes:

$$\frac{dx_i}{dt} = \mu(t) \cdot [\frac{\partial E(\mathbf{x})}{\partial x_i} + c(t) \cdot N_i(t)] \tag{1.5}$$

where $c(t)$ is called the controlling parameter and $N_i(t)$ is the noise sequence.

Algorithms for constrained optimisation. The principle of solving constrained optimisation is to convert it into unconstrained optimisation by constructing an energy function containing both cost function and constraints. After conversion, the gradient algorithms used for unconstrained optimisation can be employed to solve constrained optimisation. Two common converting methods are the penalty function and the Lagrange multiplier.

Let $f(\mathbf{x})$ be a cost function and $h(\mathbf{x}) = 0$ and $g(\mathbf{x}) \geq 0$ be equality and inequality constraints, respectively. The energy function can be defined as:

$$E(\mathbf{x}, \mathbf{k}) = f(\mathbf{x}) + P[h(\mathbf{x}), g(\mathbf{x}), \mathbf{k}] \tag{1.6}$$

where $\mathbf{k} = [k_1, k_2, ..., k_m]^T$ is a vector of controlling parameters and $P[h(\mathbf{x}), g(\mathbf{x}), \mathbf{k}]$ is a real-valued non-negative function called the penalty function which is defined as:

$$P[] \begin{cases} = 0 & if \ h(\mathbf{x}) = 0 \ and \ g(\mathbf{x}) \geq 0 \\ > 0 & otherwise \end{cases} \tag{1.7}$$

When the energy function $E(\mathbf{x})$ is minimised using a gradient algorithm, it can reach a local minimum if the constraints are satisfied, i.e. $P[] = 0$. Otherwise, P will give a penalty as it increases the value of $E(\mathbf{x})$. In the Lagrange multiplier method, the energy function is formed as:

$$L(\mathbf{x}, \lambda) = f(\mathbf{x}) + \lambda \cdot (h(\mathbf{x}) + g(\mathbf{x})) \tag{1.8}$$

where the components of the vector $\lambda = [\lambda_i, \lambda_2, \cdots]$ are the Lagrange multipliers. $L(\mathbf{x}, \lambda)$ can be minimised by employing any gradient algorithm to solve the following system of differential equations:

$$\frac{d\mathbf{x}}{dt} = -\mu \cdot \nabla_{\mathbf{x}} L(\mathbf{x}, \lambda) \tag{1.9a}$$

$$\frac{d\lambda}{dt} = -\mu \cdot \nabla_{\lambda} L(\mathbf{x}, \lambda) \tag{1.9b}$$

1.4.5 Example Neural Networks

This section briefly describes some example neural networks and associated learning algorithms.

Multi-Layer Perceptron (MLP). MLPs are perhaps the best known feedforward networks. Figure 1.9 shows an MLP with three layers: an input layer, an output layer and an intermediate or hidden layer.

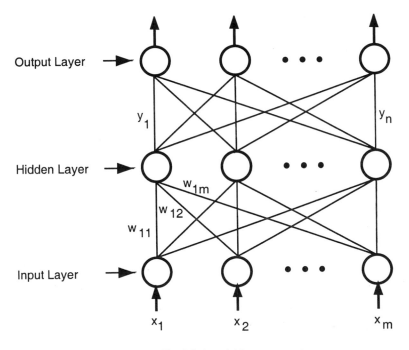

Fig. 1.9 A multi-layer perceptron

Neurons in the input layer only act as buffers for distributing the input signals x_i to neurons in the hidden layer. Each neuron j in the hidden layer sums up its input signals x_i after weighting them with the strengths of the respective connections w_{ji} from the input layer and computes its output y_j as a function f of the sum, viz.:

$$y_j = f\left(\Sigma \, w_{ji} x_i\right) \tag{1.10}$$

f can be a simple threshold function, a sigmoidal or a hyperbolic tangent function (see Table 1.2). The output of neurons in the output layer is computed similarly.

The backpropagation (BP) algorithm, a gradient descent algorithm, is the most commonly adopted MLP training algorithm. It gives the change Δw_{ji} in the weight of a connection between neurons i and j as follows:

$$\Delta w_{ji} = \eta \, \delta_j \, x_i \tag{1.11}$$

where η is a parameter called the learning rate and δ_j is a factor depending on whether neuron j is an output neuron or a hidden neuron. For output neurons,

$$\delta_j = \frac{\partial f}{\partial net_j}\left(y_j^{(t)} - y_j\right) \tag{1.12}$$

and for hidden neurons,

$$\delta_j = \frac{\partial f}{\partial net_j}\left(\sum_q w_{qj} \, \delta_q\right) \tag{1.13}$$

In equation (1.12), net_j is the total weighted sum of input signals to neuron j and $y_j^{(t)}$ is the target output for neuron j.

As there are no target outputs for hidden neurons, in equation (1.13), the difference between the target and actual outputs of a hidden neuron j is replaced by the weighted sum of the δ_q terms already obtained for neurons q connected to the output of j. Thus, iteratively, beginning with the output layer, the δ term is computed for neurons in all layers and weight updates determined for all connections.

In order to improve the training time of the backpropagation algorithm, a term called the momentum term which is proportional to the amount of the previous weight change is also added to the weight change given by equation (1.11):

$$\Delta w_{ji}(k+1) = \eta \delta_j x_i + \alpha \Delta w_{ji}(k) \tag{1.14}$$

where $\Delta w_{ji}(k+1)$ and $\Delta w_{ji}(k)$ are the weight changes after the adjustment and before the adjustment respectively and α is the momentum coefficient.

Another learning algorithm suitable for training MLPs is the GA introduced in Section 1.1.

Hopfield Network. Figure 1.10 shows one version of a Hopfield network. This network normally accepts binary and bipolar inputs (+1 or −1). It has a single "layer" of neurons, each connected to all the others, giving it a recurrent structure, as mentioned earlier. The "training" of a Hopfield network takes only one step, the weights w_{ij} of the network being assigned directly as follows:

$$w_{ij} = \begin{cases} \dfrac{1}{N} \sum\limits_{c=1}^{P} x_i^c x_j^c, & i \neq j \\ 0, & i = j \end{cases} \tag{1.15}$$

where w_{ij} is the connection weight from neuron i to neuron j, x_i^c (which is either +1 or −1) is the ith component of the training input pattern for class c, P is the number of classes and N is the number of neurons (or the number of components in the input pattern). Note from equation (1.15) that $w_{ij}=w_{ji}$ and $w_{ii}=0$, a set of conditions that guarantees the stability of the network. When an unknown pattern is input to the network, its outputs are initially set equal to the components of the unknown pattern, viz.:

$$y_i(0) = x_i \qquad 1 \leq i \leq N \tag{1.16}$$

Starting with these initial values, the network iterates according to the following equation until it reaches a minimum "energy" state, i.e. its outputs stabilise to constant values:

$$y_i(k+1) = f\left[\sum_{j=1}^{N} w_{ij} y_i(k) \right] \quad 1 \leq i \leq N \tag{1.17}$$

where f is a hard limiting function defined as:

$$f(x) = \begin{cases} -1, & x < 0 \\ 1, & x \geq 0 \end{cases} \qquad (1.18)$$

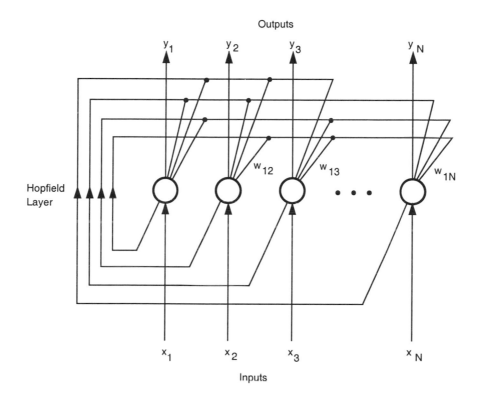

Fig. 1.10 A Hopfield network

Boltzmann Machines. Hopfield networks suffer from a tendency to stabilise to a local minimum rather than a global minimum of the energy function. This problem is overcome by a class of neural networks known as Boltzmann machines [Aarts and Korst, 1989]. These networks consist of neurons that change their states in a statistical manner rather than a deterministic way. There is a close analogy between the way these neurons operate and the way in which a solid is annealed (see also simulated annealing in Section 1.3).

If the rules employed for a state change in the binary Hopfield network are described statistically rather than deterministically, a Boltzmann machine is produced. To accomplish this, the probability of a weight change is defined in terms of the amount by which the *net* output of a neuron exceeds its threshold level, which is expressed as the following:

$$E_k = net_k - \theta_k \qquad\qquad (1.19)$$

where net_k is the *net* output of neuron k, θ_k is the threshold of neuron k and E_k is the artificial energy of neuron k. The probability is then given by:

$$P_k = 1/[1 + exp(-\Delta E_k/T)] \qquad\qquad (1.20)$$

where T is the artificial temperature and ΔE_k is the change in the energy of neuron k. Note that the Boltzmann probability function characterising the probability distribution of system energies at a given temperature is in the denominator.

The network is trained as follows. First, as in the operation of ordinary simulated annealing, T is set to a high value and hence the neurons are clamped to an initial state determined by an input vector. Second, a set of inputs is applied to the network and then the outputs and the objective function are computed. Third, a random weight change is performed and the outputs and the change in the objective function due to the weight change are recalculated. If the change is negative then the weight change is retained, otherwise it is accepted according to the Metropolis criterion in equation (1.1). This process is repeated until an acceptably low value for the objective function is achieved. At this point a different input vector is applied and the training process is repeated [Wasserman, 1989].

Kohonen Network. A Kohonen network is a self-organising feature map that has two layers, an input buffer layer to receive the input pattern and an output layer (see Figure 1.11). Neurons in the output layer are usually arranged as a regular two-dimensional array. Each output neuron is connected to all input neurons. The weights of the connections form the components of the reference vector associated with the given output neuron.

Training a Kohonen network involves the following steps:

1. Initialise the reference vectors of all output neurons to small random values;
2. Present a training input pattern;
3. Determine the winning output neuron, i.e. the neuron whose reference vector is closest to the input pattern. The Euclidean distance between a reference vector and the input vector is usually adopted as the distance measure;

4. Update the reference vector of the winning neuron and those of its neighbours. These reference vectors are brought closer to the input vector. The adjustment is greatest for the reference vector of the winning neuron and decreased for reference vectors of neurons further away. The size of the neighbourhood of a neuron is reduced as training proceeds until, towards the end of training, only the reference vector of a winning neuron is adjusted.

In a well-trained Kohonen network, output neurons that are close to one another have similar reference vectors. After training, a labelling procedure is adopted where input patterns of known classes are fed to the network and class labels are assigned to output neurons that are activated by those input patterns. An output neuron is activated by an input pattern if it wins the competition against other output neurons, that is, if its reference vector is closest to the input pattern.

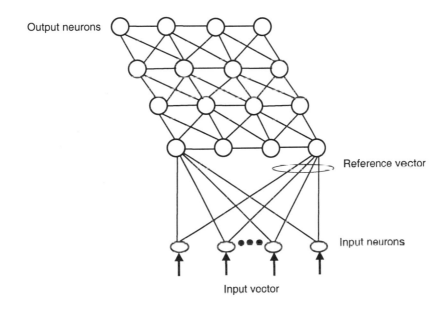

Fig. 1.11 A Kohonen network

1.5 Performance of Different Optimisation Techniques on Benchmark Test Functions

Four test functions [De Jong, 1975] were employed to measure the performance of the algorithms for numerical optimisation. The functions are listed below.

$$F_1 = \sum_{i=1}^{3} x_i^2 \tag{1.21}$$

$$F_2 = 100(x_1^2 - x_2)^2 + (1 - x_1)^2 \tag{1.22}$$

$$F_3 = \sum_{i=1}^{5} [x_i] \tag{1.23}$$

where $[x_i]$ represents the greatest integer less than or equal to x_i.

$$F_4 = [0.002 + \sum_{j=1}^{25} \frac{1}{j + \sum_{i=1}^{2}(x_i - a_{ij})^6}]^{-1} \tag{1.24}$$

where $\{(a_{1j}, a_{2j})\}_{j=1}^{25} = \{(-32, -32), (-16, -32), (0, -32), (16, -32), (32, -32),$
$(-32, -16), (-16, -16), (0, -16), (16, -16), (32, -16), \cdots, (-32, 32), (-16, 32), (0, 32), (16, 32), (32, 32)\}$.

The search spaces and resolutions for these four optimisation problems are summarised in Table 1.3.

Table 1.3 Search spaces and resolutions

Function	Parameter numbers	Solution		Parameter bounds		Resolution
		x_i	F_i	Lower	Upper	
F_1	3	0.0	0.0	-5.12	5.11	0.01
F_2	2	1.0	0.0	-2.048	2.047	0.001
F_3	5	-5.12	-30.0	-5.12	5.11	0.01
F_4	2	-32.0	1.0	-65.536	65.535	0.001

1.5.1 Genetic Algorithm Used

The GA adopted is that defined by Grefenstette [1986]. The flowchart of the algorithm is given in Figure 1.12.

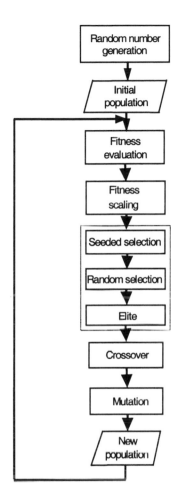

Fig. 1.12 Genetic algorithm used

Each individual in the population of solutions is sent to the evaluation unit and its fitness is calculated to determine how well it optimises the given test functions. The fitness *fit* of the individual is then evaluated from F as:

$$fit = (10000 - F) \tag{1.25}$$

The fitness scaling procedure is responsible for obtaining scaled fitness values from the raw fitness values. Fitness scaling enables better individuals to be distinguished from good individuals towards the end of the optimisation.

The selection unit implements three different procedures, random selection, seeded selection and elite selection. With random selection, a number of individuals are randomly picked to pass from one generation to the next. The percentage picked is controlled by a parameter known as the "generation gap". The seeded selection procedure simulates the biased roulette wheel technique to favour individuals with higher fitnesses. The elite procedure ensures that the fittest member of a generation passes unchanged to the next generation.

The crossover and mutation procedures perform the crossover and mutation operations described earlier.

1.5.2 Tabu Search Algorithm Used

A solution is represented as a vector, i.e. $s=(x, \sigma)$. The first vector $x(x_1,.., x_i,.., x_n)$ represents a point in the search space. The second vector $\sigma(\sigma_1,.., \sigma_i,.., \sigma_n)$ is a vector of standard deviations. A neighbour solution is produced from the present solution by replacing non-tabu x_i by

$$x_i' = x_i + N(0, \sigma_i) \tag{1.26}$$

where $N(0, \sigma_i)$ is a random Gaussian number with a mean of zero and standard deviation σ_i. In order to determine the value of σ_i, Recenberg's "1/5 success rule" is employed [Back *et al.*, 1991]. According to this rule the ratio r of successful changes to all changes made should be 1/5. The variance of the changes must be increased if r is higher than 1/5, otherwise, it must be decreased [Michalewicz, 1992]. This rule is expressed by:

$$\sigma_i' = \begin{cases} C_d\, \sigma_i, & if\ r(k) < 1/5 \\ C_i\, \sigma_i, & if\ r(k) > 1/5 \\ \sigma_i, & if\ r(k) = 1/5 \end{cases} \tag{1.27}$$

where $r(k)$ is the success ratio of the changes made during the last k iterations, and $C_i > 1$, $C_d < 1$ controls the increase and decrease rates for the variance of the changes [Michalewicz, 1992]. In the benchmarking, for C_i, C_d and k the following values were taken:

$C_i = 1.2$, $C_d = 0.6$, $k = 10$

In this work, if the number of parameters to be optimised is less than four, only the frequency-based memory is employed to classify whether a solution is tabu or not, otherwise, both the recency and frequency memories are used. This is because when the number of parameters is less than four all moves pass into the tabu list in just a few iterations.

1.5.3 Simulated Annealing Algorithm Used

The simulated annealing algorithm adopted employs a generation mechanism similar to the one used in the tabu search algorithm. The values selected for C_i and C_d are 1.1 and 0.9, respectively. k is taken as 10. The representation of solutions is in the form of a floating-point number vector. The number of neighbour solutions is equal to twice the number of parameters.

The temperature updating rule employed is the following:

$$T_{i+1} = c\, T_i \ , \ i = 0,1\ldots \tag{1.28}$$

where c is taken as 0.9.

1.5.4 Neural Network Used

For the optimisation of Function 1, Function 2 and Function 4, Hopfield neurons are employed to construct neural networks implementing the steepest-descent algorithm. This type of neural network is illustrated in Figure 1.13. The cost functions are regarded as energy functions. The variables are used as the outputs of the neurons. They are fedback to the input sides of all Hopfield neurons through partial-derivative computing components $\dfrac{\partial E(\mathbf{x})}{\partial x_i}$. For Function 1 and Function 2, the learning rates ($\mu(t)$'s) are fixed. For the last function,

$$\mu(t) = \frac{0.05}{\sqrt{(\dfrac{\partial E}{\partial x_1})^2 + (\dfrac{\partial E}{\partial x_2})^2}}$$

in order to increase the speed of convergence. For this function, a noise sequence is also added during optimisation to avoid the search process being trapped in a local minimum. This noise sequence is added as follows:

$$\tilde{E} = E + c(t)\sum_i x_i N_i(t) \tag{1.29}$$

where $c(t)$ is called a control parameter and $N_i(t)$ stands for the noise sequence. The control parameter has the form of $c(t) = \beta e^{-\alpha t}$ $(\beta = 1.0, \alpha = 1.0)$. This noise sequence is added after $t = 200$.

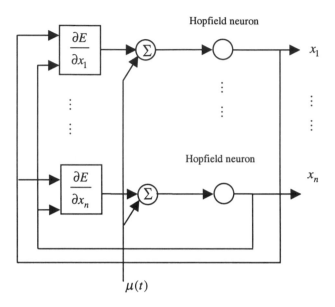

Figure 1.13. Functional block diagram of a neural network for optimisation based on steepest-descent algorithm

A standard Hopfield network (see Section 1.4.5) was used for the optimisation of Function 3. A suitable energy function was

$$E = -\frac{1}{2}\sum_{i=1}^{5}\sum_{j=1}^{5}\left(\frac{a}{2}x_i x_j\right) - \sum_{i=1}^{5}(-ax_i) \qquad (1.30)$$

where a is a scalar quantity and x_i $(i = 1, \cdots 5)$ is the output of the ith neuron. The learning rate is $\mu(t) = \dfrac{0.1}{a}$.

The genetic algorithm, tabu search (TS), simulated annealing (SA) and neural network (NN) programs which were employed in the optimisation of the test functions are included in Appendices A3, A4, A5 and A6, respectively.

1.5.5 Results

The number of evaluations was taken as equal to 10,000 for all algorithms and functions. In the case of the genetic algorithm, as the population size was 50, the generation number was equal to 200 for all functions. The other control parameters of the genetic algorithm were crossover rate = 0.85 and mutation rate = 0.015. The length of the binary strings used for the representation of a parameter depends on the resolution required. The length of a binary string employed to represent a parameter is 30, 24, 50 and 34 for Functions 1, 2, 3 and 4, respectively.

In the tabu search, the number of neighbours produced in an iteration was equal to twice the number of parameters. The length of a solution vector depends on the number of parameters of the function being optimised. Therefore, it was 3, 2, 5 and 2 for Functions 1, 2, 3 and 4, respectively. The recency and the frequency factors used were 5 and 2, respectively.

The initial temperature in the simulated annealing algorithm was chosen as 100. The temperature decreased according to the formula given in equation (1.28). At each temperature, 10 iterations were executed. The number of neighbours produced from a solution was equal to double the number of parameters.

Each algorithm ran 30 times for each function with different seed numbers to examine their robustness. The histograms drawn from the results obtained are shown in Figures 1.14-1.17. In order to give an indication of the convergence speed of the search, the values of these functions at discrete stages of the search are presented in Tables 1.4-1.7. Note that the first value seen in the GA columns in the tables belongs to the best solution among 50 initial solutions.

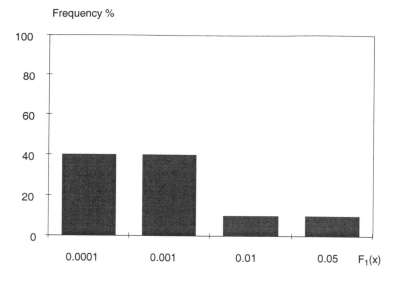

Fig. 1.14 (a) Histogram of results obtained by GA for F_1

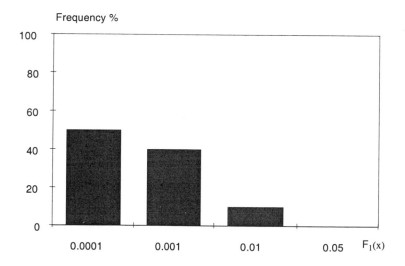

Fig. 1.14 (b) Histogram of results obtained by TS for F_1

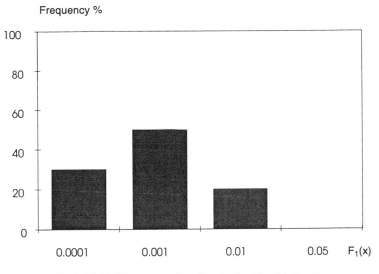

Fig. 1.14 (c) Histogram of results obtained by SA for F_1

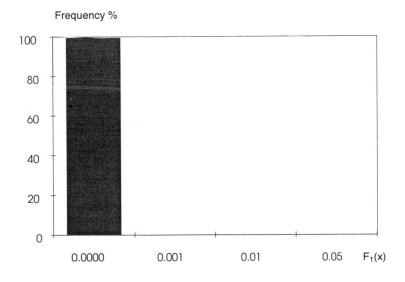

Fig. 1.14 (d) Histogram of results obtained by NN for F_1

Frequency %

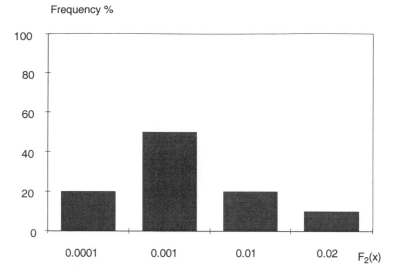

Fig. 1.15 (a) Histogram of results obtained by GA for F_2

Frequency %

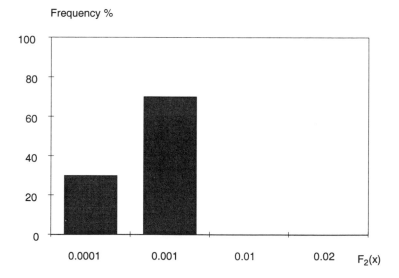

Fig. 1.15 (b) Histogram of results obtained by TS for F_2

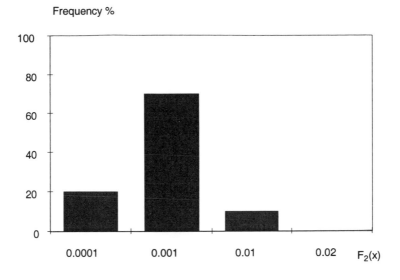

Fig. 1.15 (c) Histogram of results obtained by SA for F_2

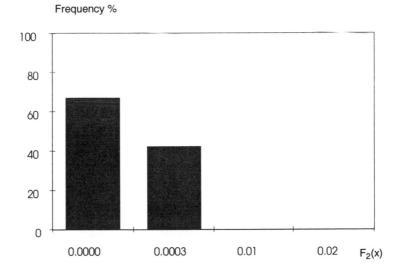

Fig. 1.15 (d) Histogram of results obtained by NN for F_2

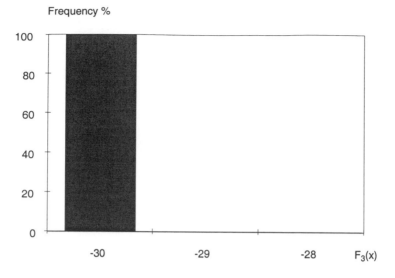

Fig. 1.16 (a) Histogram of results obtained by GA for F_3

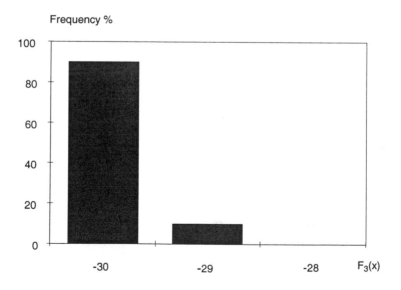

Fig. 1.16 (b) Histogram of results obtained by TS for F_3

Frequency %

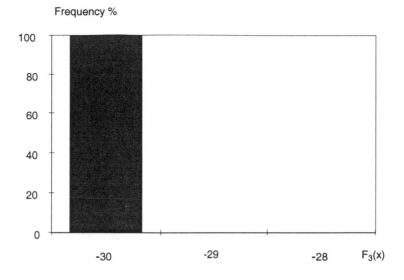

Fig. 1.16 (c) Histogram of results obtained by SA for F_3

Frequency %

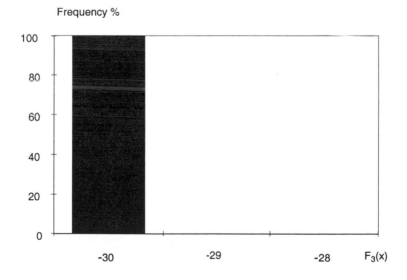

Fig. 1.16 (d) Histogram of results obtained by NN for F_3

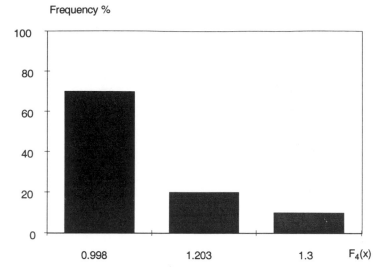

Fig. 1.17 (a) Histogram of results obtained by GA for F_4

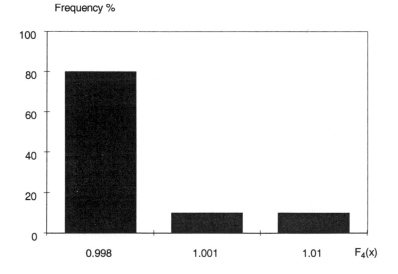

Fig. 1.17 (b) Histogram of results obtained by TS for F_4

Frequency %

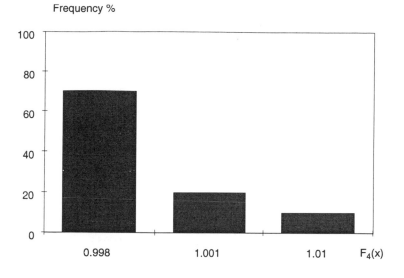

Fig. 1.17 (c) Histogram of results obtained by SA for F_4

Frequency %

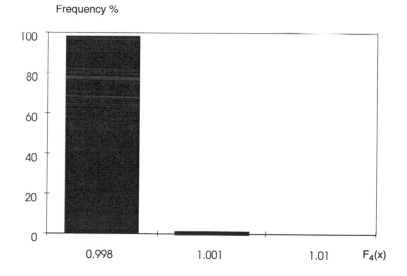

Fig. 1.17 (d) Histogram of results obtained by NN for F_4

Table 1.4 Values found by the algorithms for F_1 at discrete stages of search

Evaluation Number	Genetic Algorithm	Tabu Search	Simulated Annealing	Neural Network
0	2.5390	14.3225	12.2531	52.3219
10	*	8.3762	3.3856	52.1961
20	*	1.4065	1.1289	51.2517
30	*	0.2324	0.4237	45.1706
40	*	0.0096	0.3296	23.6667
50	2.5390	0.0078	0.3454	4.8957
100	2.5390	0.0078	0.0467	0.0001
500	2.4762	0.00014	0.0015	0.0001
1000	0.7734	0.00014	0.0015	0.0001
2000	0.0021	0.00014	0.0005	0.0001
5000	0.0006	0.00014	0.0005	0.0001
10000	0.0006	0.00014	0.0005	0.0001

Table 1.5 Values found by the algorithms for F_2 at discrete stages of search

Evaluation Number	Genetic Algorithm	Tabu Search	Simulated Annealing	Neural Network
0	3.5634	4.5067	3.6875	2480.68
10	*	0.3456	1.1214	1.1646
20	*	0.3456	1.0325	0.0297
30	*	0.3456	0.9889	0.0021
40	*	0.3456	0.9978	0.0006
50	3.5634	0.3456	0.6582	0.0004
100	1.5332	0.3456	0.5392	0.0003
500	0.1945	0.3078	0.1075	0.0003
1000	0.0031	0.0586	0.0877	0.0003
2000	0.0025	0.0017	0.0877	0.0003
5000	0.0025	0.0009	0.0019	0.0003
10000	0.0025	0.0005	0.0019	0.0003

Table 1.6 Values found by the algorithms for F_3 at discrete stages of search

Evaluation Number	Genetic Algorithm	Tabu Search	Simulated Annealing	Neural Network
0	-16	-1	-10	-3
10	*	-6	-12	-7
20	*	-11	-13	-11
30	*	-19	-18	-16
40	*	-20	-19	-30
50	-16	-22	-26	-30
100	-19	-27	-28	-30
500	-27	-30	-30	-30
1000	-27	-30	-30	-30
2000	-28	-30	-30	-30
5000	-30	-30	-30	-30
10000	-30	-30	-30	-30

Table 1.7 Values found by the algorithms for F_4 at discrete stages of search

Evaluation Number	Genetic Algorithm	Tabu Search	Simulated Annealing	Neural Network
0	3.8756	499.1245	421.6523	499.9998
10	*	261.3594	125.7685	499.9998
20	*	56.3578	92.4536	499.9997
30	*	54.9811	68.2361	499.9997
40	*	51.2734	12.6724	499.9997
50	3.8756	19.5573	7.8743	499.9997
100	3.8756	10.1763	2.0643	499.9949
500	3.7965	1.1101	2.0643	499.9991
1000	3.7965	0.9980	0.9983	0.9991
2000	2.9922	0.9980	0.9983	0.9991
5000	0.9801	0.9980	0.9983	0.9991
10000	0.9980	0.9980	0.9981	0.9991

1.6 Performance of Different Optimisation Techniques on Travelling Salesman Problem

The Travelling Salesman Problem (TSP) is a classic combinatorial optimisation problem. In this problem, a travelling salesman is expected to visit a set of cities A, B, C, L . The objective of this problem is to find the shortest tour in which the travelling salesman visits each city once and returns to the starting city. Genetic algorithms, tabu search, simulated annealing, and neural networks were applied to a TSP consisting of 50 cities to evaluate their performances. The map of these 50 cities was created randomly in a unit square. This map is shown in Figure 1.18.

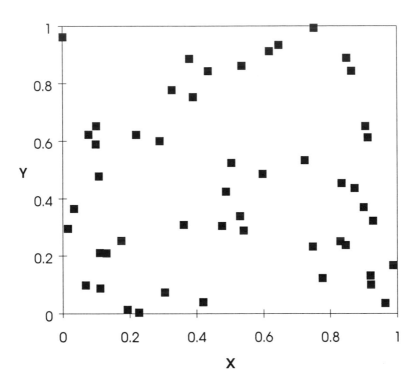

Figure 1.18 City map of 50 cities

1.6.1 Genetic Algorithm Used

The genetic algorithm employed represents the 50-city TSP with a permutation list $P(C_1, C_2, L\ C_{50})$, where C_i is a binary string. C_i stands for a city and the subscript is the index of the city. Every city can be arranged into all the positions of the list. The "Manhattan" distance between two neighbourhood cities, denoted as $d_{c_i c_{i+1}}$, is called the inter-city distance. An arrangement of 50 cities on the permutation list represents a tour. The length of the tour is the sum of all inter-city distances in the list. Each arrangement is used as an individual and the tour length function is the cost function F in the genetic algorithm. The constraint of visiting each city once is ensured because every city must appear on the list only once.

The algorithm started with 64 initial individuals. They were then divided into several sub-populations and the fitness for every individual in all sub-populations was calculated. At the next step, the algorithm selected the first 8 best sub-populations as parents for the successive population. The proportions for these sub-populations are 0.25:0.20:0.15:0.10:0.10:0.10:0.05:0.05. In the last step, the algorithm deposits the individuals obtained from the last step into a mating pool and applies crossover and mutation operations. The string size is randomly chosen in the interval $[\frac{N}{4}, \frac{N}{2}]$ and the mutation rate is 0.01 [Muhlenbein *et al.*, 1988; Martina, 1989].

1.6.2 Tabu Search Algorithm Used

The tabu search algorithm also encodes the TSP with a permutation list. The values of the list entry are of 1 to 50, which are the indices of the 50 cities in the city map. An arrangement of the index numbers on the list stands for a tour. An edge is defined as the "Manhattan" distance between two cities. The length of a tour is the sum of all edges in the list. The tour distance function is also the cost function of the TSP. Each index number must appear in the list once.

A tabu list embodies the edges of previous moves. The tabu list needs to be examined in order to ensure that a new move is not tabu from the present solution. A new move is created by removing an index number from its position in the permutation list and by inserting it between two other index numbers. If the new edges created by the insertion are not in the tabu list the move is accepted as a new move.

Both recency and frequency are used to determine the tabu list. The parameters for recency and frequency were $r = 0.5$ and $f = 1.5$, respectively.

1.6.3 Simulated Annealing Algorithm Used

The TSP is encoded in the simulated annealing algorithm in the same way as it is in the tabu search algorithm.

In simulated annealing, the rearrangements of the index numbers 1 to 50 on the list result from temperature updating. The temperature updating is determined by an annealing schedule. The annealing schedule and the parameters are as follows [Kirkpatrick, 1983]:

1. Let the initial temperature $T_0 = 10$, and the length of time $L_0 = 50$;

2. Heat up by updating temperature to $T = T_0 / 0.8$ and setting $L = 50$ until the variance of the cost function < 0.05;

3. Cool down by updating temperature to $T = 0.95T$ and setting $L = 50$ when the percentage of the accepted arrangement $> 50\%$;

4. Cool down slowly by updating temperature to $T = 0.95T$ and updating L to $16L$ when there is no accepted arrangement.

1.6.4 Neural Network Used

The 50-city TSP is mapped onto a 50×50 array. The row index of the array represents cities, and the column index indicates the position of the travelling salesman. The entries of the array can be either 1 or 0 representing a city being visited or not, respectively. The constraint of visiting each city once means there is a single 1-value entry in every row and every column. For 50 cities, a total of 50 groups of 50 neurons are needed to represent the 50×50 array. Each neuron is viewed as an entry in the array [Zurada, 1992].

Hopfield networks are suitable for solving small size TSPs consisting of 10 cities or fewer [Hopfield and Tank, 1985]. For a problem involving more than 10 cities, a Potts neural network [Carster and Bo, 1989; Carster, 1990] has been shown to produce good results and is the one reported here. A Potts neural network utilises a multidimensional activation function:

$$v_{ia} = \frac{e^{u_{ia}}}{\sum_b e^{u_{ia}}} \tag{1.31}$$

where u_{ia} is the internal potential of the ith neuron, which is defined as:

$$u_{ia} = -\frac{\partial E}{\partial v_{ia}} \tag{1.32}$$

Due to this activation function, the outputs of any group of 50 neurons always satisfy the relation $\sum_a v_{ia} = 1$. The energy function used in the Potts neural network for solving the given TSP can thus be defined as follows:

$$E = \sum_i \sum_j d_{ij} \sum_a v_{ia} v_{j(a+1)} - \sum_i \sum_a v_{ia}^2 + \sum_a (\sum_i v_{ia})^2 \tag{1.33}$$

where d_{ij} is the distance between cities i and j.

1.6.5 Results

The overall performances of the four algorithms are listed in Table 1.8.

Table 1.8 Tour lengths obtained

Genetic algorithm	Tabu search	Simulated annealing	Neural network
5.58	5.68	6.80	6.61

1.7 Summary

This chapter introduced the basic principles of four artificial intelligence-based optimisation techniques: genetic algorithms, tabu search, simulated annealing algorithms and neural networks. The optimisation of a set of numerical test functions was used to demonstrate the performance of each technique. A 50-city travelling salesman problem was also employed to test their performances on a combinatorial optimisation task.

References

Aarts, E. and Korst, J. (1989) *Simulated Annealing and Boltzmann Machines*, Wiley, Chichester, UK.

Back, T., Hoffmeister, F. and Schwefel, H.P. (1991) A survey of evolution strategies, *Proc. 4th Int. Conf. on Genetic Algorithms and their Applications*, Morgan Kaufmann, San Mateo, CA, pp.2-9.

Baker, J.E. (1985) Adaptive selection methods for genetic algorithms, *Proc. 1st Int. Conf. on Genetic Alorithms and their Applications*, Lawrence Erlbaum Associates, Hillsdale, NJ, pp.101-111.

Carpenter, G. A. and Grossberg, S. (1988) The ART of Adaptive Pattern Recognition by a Self-Organising Neural Network, *Computer*, March, pp.77-88.

Carster, P. and Bo, S. (1989) A new method for mapping optimization problems onto neural networks, *Int. J. of Neural Systems*, Vol.1, No.1, pp.3-22.

Carster, P. (1990) Parallel distributed approaches to combinatorial optimization: benchmark studies on traveling salesman problem, *Neural Computation*, Vol.2, pp.261-269.

Cichocki, A., and Unbehauen, R. (1993) *Neural Network for Optimization Signal Processing*, Wiley, Chichester, UK.

Davis, L. (1991) *Handbook of Genetic Algorithms*, Van Nostrand Reinhold, New York, NY.

De Jong, K.A. (1975) *Analysis of the Behavior of a Class of Genetic Adaptive Systems*, Ph.D. Dissertation, Department of Computer and Communication Science, University of Michigan, Ann Arbor, MI.

Elman, J. L. (1990) Finding structure in time, *Cognitive Science*, Vol.14, pp.179-211.

Fogarty, T.C. (1989) Varying the probability of mutation in the genetic algorithm, *Proc. 3rd Int. Conf. on Genetic Algorithms and Their Applications*, George Mason University, pp.104-109.

Glover, F. (1986) Future paths for integer programming and links to artificial intelligence, *Computers and Operation Research*, Vol.13, pp.533-549.

Glover, F. (1989) Tabu Search-part I, *ORSA Journal on Computing*, Vol.1., pp.190-206.

Glover, F. (1990) Tabu Search-part II, *ORSA Journal on Computing*, Vol.2., pp.14-32.

Goldberg, D. E. (1989) *Genetic Algorithms in Search, Optimisation and Machine Learning,* Addison-Wesley, Reading, MA.

Grefenstette, J. J. (1986) Optimisation of control parameters for genetic algorithms, *IEEE Trans. on Systems, Man and Cybernetics*, Vol.SMC-16, No.1, pp.122-128.

Hansen, P. (1986) The steepest ascent mildest descent heuristic for combinatorial programming, *Conf. on Numerical Methods in Combinatorial Optimisation*, Capri, Italy.

Hecht-Nielsen, R. (1990) *Neurocomputing,* Addison-Wesley, Reading, MA.

Holland, J. H. (1975) *Adaptation in Natural and Artificial Systems*, University of Michigan Press, Ann Arbor, MI.

Hopfield, J. J. (1982) Neural networks and physical systems with emergent collective computational abilities, *Proc. National Academy of Sciences*, USA, April, Vol.79, pp.2554-2558.

Hopfield, J. J., and Tank, D. W. (1985) "Neural" computation of decisions in optimization problems, *Biol. Cybern.*, Vol.52, pp.141-152.

Jordan, M. I. (1986) Attractor dynamics and parallelism in a connectionist sequential machine, *Proc. 8th Annual Conf. of the Cognitive Science Society*, Amherst, MA, pp.531-546.

Kirkpatrick, S., Gelatt, C.D. Jr and Vecchi, M.P. (1983) Optimization by simulated annealing, *Science*, Vol.220, No.4598, pp.671-680.

Kohonen, T. (1989) *Self-Organisation and Associative Memory* (3rd ed.), Springer-Verlag, Berlin.

Martina, G. S. (1989) ASPARAGOS An asynchronous parallel genetic optimization strategy, *Proc. 3rd Int. ICGA Conf.*, pp.422-427.

Metropolis, N., Rosenbluth, A.W., Rosenbluth, M.N., Teller, A.H. and Teller, E. (1953) Equation of state calculations by fast computing machines, *J. of Chem. Phys.*, Vol.21, No.6, pp.1087-1092.

Michalewicz, Z. (1992) *Genetic Algorithms + Data Structures = Evolution Programs*, Springer-Verlag, New York, NY.

Muhlenbein, H., Gorges-Schleuter, M. and Kramar, O. (1988) Evolution algorithms in combinatorial optimization, *Parallel Computing*, Vol.7, pp.65-85.

Osman, I.H. (1991) *Metastrategy Simulated Annealing and Tabu Search Algorithms for Combinatorial Optimisation Problems*, PhD Thesis, University of London, Imperial College, UK.

Reeves, C.R. (1995) *Modern Heuristic Techniques for Combinatorial Problems*, McGraw-Hill, Maidenhead, UK.

Rumelhart, D. E. and J. L. McClelland (1986) *Parallel Distributed Processing: Explorations in the Microstructure of Cognition,* MIT Press, Cambridge, MA.

Schaffer, J.D., Caruana, R.A., Eshelman, L.J. and Das, R. (1989) A study of control parameters affecting on-line performance of genetic algorithms for function optimisation, *Proc. 3rd Int. Conf. on Genetic Algorithms and their Applications*, George Mason University, pp.51-61.

Wasserman, P.D. (1989) *Neural Computing*, Van Nostrand Reinhold, New York, NY.

Whitely, D. and Hanson, T. (1989) Optimising neural networks using faster, more accurate genetic search, *Proc. 3rd Int. Conf. on Genetic Algorithms and their Applications*, George Mason University, pp.370-374.

Widrow, B. and M. E. Hoff (1960) Adaptive switching circuits, *Proc. IRE WESCON Convention Record*, Part 4, IRE, New York, pp.96-104.

Zurada, J. M. (1992) *Introduction to Artificial Neural Systems*, West, Eagan, MN.

Chapter 2

Genetic Algorithms

This chapter consists of two main sections. The first section describes four new models for genetic algorithms. The second section presents applications of genetic algorithms to problems from different area engineering.

2.1 New Models

This section presents four new modified genetic algorithms. The first is a version of micro genetic algorithms, called a hybrid genetic algorithm, which uses a modified reproduction mechanism [Pham and Jin, 1996]. The second model is based on the structure of parallel genetic algorithms [Pham and Karaboga, 1991b], utilising cross-breeding for genetic optimisation. The third model employs four techniques to help a GA to find multiple, alternative solutions for a problem [Pham and Yang, 1993]. The last genetic algorithm uses three simple strategies for automatically adapting the mutation rate [Pham and Karaboga, 1997].

2.1.1 Hybrid Genetic Algorithm

Currently, a number of GAs used within the GA community employ a reproduction operator based on roulette wheel selection. If such GAs deal with a small population, their performance becomes poor because, by the nature of roulette wheel selection, the population may lose genetic diversity at an early stage due to the dominance of fit individuals. Therefore, a rule of thumb is to use a population size ranging from 30 to 200 [Krishnakumar, 1989]. A GA with a population within this range drastically increases computational burden due to the increase in the number of parameters to be treated. This limits the use of the GA to off-line applications. This fact motivates this investigation into a new reproduction operator that can overcome shortcomings of the reproduction operator based on roulette wheel selection.

A new version of the reproduction operator has been developed and that reproduction operator in conjunction with the simple crossover and mutation operators has been implemented in the proposed hybrid genetic algorithm (HGA). Like many other GAs, the HGA deals with a population of binary encoded structures and begins with a randomly generated population. This population is iteratively evolved by applying these three operators, whilst the solution is being searched for.

New Reproduction Operator. Reproduction is one of the most important operations in GAs. A common method of reproduction is that based on selection. Individuals in the current population are selected on the basis of their fitnesses to form a mating pool. Individuals having higher fitnesses are more likely to be reproduced. By doing this, the reproduction operator drives the search towards better solutions. As mentioned in the previous chapter, there are several selection schemes, including roulette wheel selection [Holland, 1975; Goldberg, 1989a], tournament selection [Goldberg, 1989a] and ranking-based selection [Baker, 1985; Whitely, 1989]. Amongst the schemes available, roulette wheel selection is the most popular method. However, this type of selection has the following drawbacks:

- The fittest individual may be lost during the selection process due to its stochastic nature.
- Fit individuals may be copied several times and a fit individual may quickly dominate the population at an early stage, especially if the population size is small.
- The selection operation alone explores no new points in a search space. In other words, it cannot create new schemata.

A new reproduction operator is introduced to resolve these drawbacks. Consider an optimisation problem to which GAs are universally applicable, formulated as follows:

$$\text{Minimise } \{ F(\mathbf{x}), \mathbf{x} = [x_1 \cdots x_M]^T \in \mathbf{X}^M, \ \mathbf{X}^M \subseteq \mathfrak{R}^M \} \tag{2.1}$$

$$\text{subject to a search space } \mathbf{X}^M = \{ \mathbf{x} | \ x_j^{(L)} \leq \mathbf{x} \leq x_j^{(U)} \quad (1 \leq j \leq M) \} \tag{2.2}$$

where $F: \mathbf{X}^M \rightarrow \mathfrak{R}$ is a function, M is the parameter number and $x_j^{(L)}$ and $x_j^{(U)}$ are the lower and upper bounds of the jth parameter x_j, respectively.

Let $\mathbf{P}(k)$ be a population of N structures (individuals) $\mathbf{s}_i(k)$ ($1 \leq i \leq N$) in generation k. The new reproduction operation is carried out in the following three steps.

Step 1. N structures $\mathbf{s}_i(k)$ ($1 \leq i \leq N$) in $\mathbf{P}(k)$ are decoded by a decoding function θ^{-1} as:

$$\mathbf{x}_i(k) = [x_{i1}(k) \; x_{i2}(k) \cdots x_{iM}(k)]^T \in \mathbf{X}^M \quad (1 \leq i \leq N) \tag{2.3}$$

where $x_{ij}(k) = \theta^{-1}(s_{ij}(k))$ ($1 \leq j \leq M$) and $s_{ij}(k)$ is the jth string in $\mathbf{s}_i(k)$. Let $f_i(k)$ be the fitness of the ith structure defined as:

$$f_i(k) = -F(\mathbf{x}_i(k)) + F_{min} \quad (\geq 0) \tag{2.4}$$

where F_{min} is a coefficient chosen to ensure that the fitness is non-negative. Let $\mathbf{x}_b(k)$ be the best parameter vector that has the largest fitness, denoted $f_b(k) = \max\limits_{1 \leq i \leq N} [f_i(k)]$ (> 0), amongst structures in $\mathbf{P}(k)$.

Step 2. New values are assigned to the parameters of each individual using the following formula:

$$\bar{x}_{ij}(k+1) = x_{ij}(k) + \eta_i \left[(f_b(k)\text{-}f_i(k)) / f_b(k)\right] [x_{bj}(k) - x_{ij}(k)] \quad (1 \leq i \leq N, 1 \leq j \leq M) \tag{2.5}$$

where $x_{ij}(k)$ and $x_{bj}(k)$, respectively, are the jth parameter of $\mathbf{x}_i(k)$ and $\mathbf{x}_b(k)$, $\bar{x}_{ij}(k+1)$ is the jth new parameter of $\mathbf{x}_i(k)$ and η_i is a positive coefficient which is chosen empirically such that a reasonable improvement of the HGA performance is obtained. It should be noted that the new point is prevented from lying outside the search space by the upper bound or lower bound in equation (2.2). By letting $\bar{\mathbf{x}}_i = [\bar{x}_{i1} \cdots \bar{x}_{iM}]^T$ be a new parameter vector in $\mathbf{X}^M$, equation (2.5) can be rewritten in a compact form as:

$$\bar{\mathbf{x}}_i(k+1) = \begin{pmatrix} \sigma_i(k)x_{i1}(k) + \xi_i(k)x_{b1}(k) \\ \vdots \\ \sigma_i(k)x_{iM}(k) + \xi_i(k)x_{bM}(k) \end{pmatrix}$$

$$= \sigma_i(k)\mathbf{x}_i(k) + \xi_i(k)\mathbf{x}_b(k) \quad (1 \leq i \leq N) \tag{2.6a}$$

where

$$\sigma_i(k) = 1 - \xi_i(k) \text{ and } \xi_i(k) = \eta_i[(f_b(k)-f_i(k)) / f_b(k)] \tag{2.6b}$$

In order to show the operation graphically, two-dimensional geometric representations of $\overline{x}_i(k+1)$ are sketched in Figures 2.1 and 2.2 for two different values of η_i. A choice of η_i between 0 and 1 will assign $x_i(k+1)$ a point on the line vector $x_b(k)-x_i(k)$, whilst a choice of η_i between 1 and 2 will assign a point on the line vector $2[x_b(k)-x_i(k)]$. Note that the latter vector is twice as long as the former vector.

Step 3. N parameter vectors $\overline{x}_i(k+1)$ $(1 \leq i \leq N)$ are then encoded into structures $\overline{s}_i(k+1)$ to constitute a mating pool $\overline{P}(k+1)$ for the subsequent genetic operation.

Figure 2.3 shows the flow diagram of the new reproduction operation of the HGA. As seen in equation (2.5), the new reproduction operator assigns each individual in the current population a new parameter vector on the basis of both the normalised fitness difference and the parameter difference between the individual and the best structure. The underlying idea of this operator is that it guides the population towards the best structure. Weak individuals far from the best solution undergo more correction than strong individuals close to the best, whilst duplication of individuals is minimised. During this operation, only individuals having the same fitness as the best structure are copied. Weak individuals, which would be eliminated in roulette wheel selection, have a chance to be allocated to a point in the vicinity of the best through proper choices of η_i's. On the other hand, strong individuals as competitors of the best solution are allowed to remain near their original points with a small movement towards that solution.

One way of enhancing the performance of a GA is to help the survival of the fittest (best) individual from one generation to the next. As can be easily verified in equation (2.6), the fittest individual is automatically transferred into the mating pool. For example, if the ith individual $x_i(k)$ is the fittest, namely $f_i(k) = f_b(k)$, then from equation (2.6) $\overline{x}_i(k+1) = x_i(k)$. As the proposed operator guarantees the survival of the fittest individual during reproduction, it is apparent that the probability of that individual later surviving into the next generation is higher than that for the reproduction operator based on roulette wheel selection.

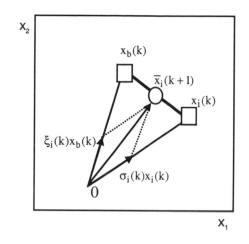

Fig. 2.1 Two-dimensional geometric representation of $\overline{\mathbf{x}}_i\,(k+1)\,(0<\eta_i\le 1)$

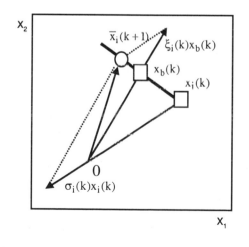

Fig. 2.2 Two-dimensional geometric representation of $\overline{\mathbf{x}}_i\,(k+1)\,(1<\eta_i\le 2)$

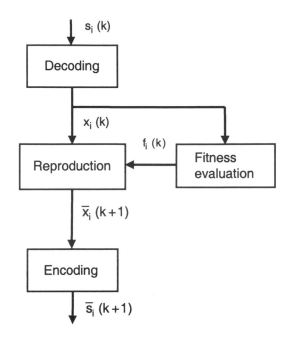

Fig. 2.3 New reproduction operation of the HGA

As already mentioned, one common problem which roulette wheel based GAs face is the possibility of premature convergence to unwanted local optima due to the predominance of fit structures. This possibility occurs mainly when the roulette wheel selection method imposes structural similarity in early generations and, due to this, the population loses genetic diversity. The greater the number of multiple copies of structures, the less genetic diversity the population has. A method of decreasing the possibility of premature convergence is to keep the numbers of multiple replications of individuals as small as possible, whilst further strengthening the population. Recall that, through the proposed reproduction operator with appropriate choices of η_i's, only individuals having the same fitness as the best structure are replicated and the other individuals are allocated to points different from themselves whilst the reproduction operator drives the population towards the best individual. This property seems to make the proposed operator a feasible method which maintains genetic diversity in early generations.

Another important property of the new reproduction operator is that it has the ability to search for new points through allocation of individuals to vector locations different from themselves. In other words, it can cause the alteration of features. Notice that, in the basic GA, only the mutation operator allows this. Owing to this property, the HGA can generate new schemata which do not belong to their parents as long as all individuals do not have the same fitness. The following theorem concerns the creation of new schemata.

Theorem 1. Let $\mathbf{P}(k)$ be a population of N structures $s_i(k)$ $(1 \leq i \leq N)$ at generation k and $f_i(k)$ be their corresponding fitness values. If fitnesses $f_i(k)$ $(1 \leq i \leq N)$ of all individuals are not the same, then the reproduction operator given by equation (2.6) can create at least one new schema which does not exist in $\mathbf{P}(k)$.

Proof. This proof will be restricted to showing that there is a proper choice of η_i with which the reproduction operator can reproduce a structure which differs from its parents because that structure itself contains a new schema. From this assumption, there is at least one individual whose fitness is different from that of the others in $\mathbf{P}(k)$. Let $s_i(k)$ be one amongst those individuals which are not the best $s_b(k)$. Thus $f_i(k) \neq f_b(k)$.

Subtracting $\mathbf{x}_j(k)$ $(1 \leq j \leq N)$ from the reproduced parameter vector of $\mathbf{x}_i(k)$ yields:

$$\overline{\mathbf{x}}_i(k+1) - \mathbf{x}_j(k) - \sigma_i(k)\mathbf{x}_i(k) + \xi_i(k)\mathbf{x}_b(k) - \mathbf{x}_j(k)$$
$$= [\mathbf{x}_i(k) - \mathbf{x}_j(k)] + \eta_i [(f_b(k) - f_i(k)) / f_b(k)] [\mathbf{x}_b(k) - \mathbf{x}_i(k)]$$

Note that $f_i(k) \neq f_b(k)$ implies $\mathbf{x}_i(k) \neq \mathbf{x}_b(k)$. If η_i is selected such that

$$\eta_i \neq \frac{f_b(k)}{[f_b(k) - f_i(k)]} \cdot \frac{[\mathbf{x}_b(k) - \mathbf{x}_i(k)]^T [\mathbf{x}_j(k) - \mathbf{x}_i(k)]}{[\mathbf{x}_b(k) - \mathbf{x}_i(k)]^T [\mathbf{x}_b(k) - \mathbf{x}_i(k)]}$$

then, $\overline{\mathbf{x}}_i(k+1) \neq \mathbf{x}_j(k)$ $(1 \leq j \leq N)$. This implies that $\overline{s}_i(k+1) \neq s_j(k)$ $(1 \leq j \leq N)$. Hence, the reproduction operator generates at least one new schema which differs from all possible schemata in $\mathbf{P}(k)$. □

For the successful operation of the overall search procedure using this reproduction operator, new parameter points must not attempt to grow outside the search space during reproduction. Therefore, a proper choice of η_i's is important to ensure the stability of the reproduction operator. Recall that the population in the HGA evolves through the three basic operators. As long as $\overline{\mathbf{x}}_i(k+1)$ is subsequently modified by the crossover and mutation operators, equation (2.6) works as a kind of static system and the boundedness of both the inputs $\mathbf{x}_i(k)$ and $\mathbf{x}_b(k)$ guarantees that of the output $\overline{\mathbf{x}}_i(k+1)$. However, there may be some situations when both the crossover and mutation operators rarely disrupt structures for a certain number of generations. This can occur if the population becomes clustered around a point after successive generations because structures so resemble each other and where often the mutation rate is very small. In this case, the reproduction operator alone is iteratively used for a certain number of generations and equation (2.6) works as a dynamic system described by a set of N independent difference equations with state vector $x_i(k)$ and input vector $x_b(k)$. The stability of these equations must be proved.

Substituting $\bar{x}_i(k+1)$ with $x_i(k+1)$ in equation (2.6), the solution of each difference equation is given by:

$$x_i(k) = \Phi_i(k, n)x_i(n) + \sum_{j=n+1}^{k} \Phi_i(k, j)\,\xi_i(j-1)x_b(j-1) \quad (1 \le i \le N) \tag{2.7}$$

where $x_i(n)$ is the initial state vector in the nth iteration and $\Phi_i(k, j)$ is the fundamental function satisfying the following equation:

$$\Phi_i(k+1, j) = \sigma_i(k)\Phi_i(k, j) \text{ for all } k \ge j \text{ and } \Phi_i(j, j) = 1 \tag{2.8}$$

Notice that equation (2.7) can be decomposed into the zero-input response (homogeneous solution) $\Phi_i(k, n)x_i(n)$ and the zero-state response (particular solution):

$$\sum_{j=n+1}^{k} \Phi_i(k, j)\,\xi_i(j-1)x_b(j-1).$$

The stability of the reproduction operator is proved in the following theorem.

Definition 1. (Stability) A linear dynamic equation is said to be stable if and only if, for all initial states and for any bounded inputs, all the state variables are bounded (see [Chen, 1984]).

Theorem 2. Let $P(k)$ be a population of N structures $s_i(k)$ $(1 \le i \le N)$ at generation k. If $P(k)$ evolves iteratively through only the reproduction operator for a certain number of generations and the search space is bounded, there are some η_i's between 0 and 2 which stabilise the reproduction operator.

Proof. First, it will be shown that $0 < \eta_i \le 2$ satisfies $|\sigma_i(k)| \le 1$ for any n and all $k \ge n$. Note that

$0 \le [(f_b(k)-f_i(k)) / f_b(k)] \le 1$ for all k

Thus

$0 \le \eta_i[(f_b(k)-f_i(k)) / f_b(k)] = \xi_i(k) = 1 - \sigma_i(k) \le 2[(f_b(k)-f_i(k)) / f_b(k)] \le 2$

Hence, $0 < \eta_i \le 2$ satisfies $|\sigma_i(k)| \le 1$ for all $k \ge n$. Notice that if the search space is bounded, then there exists a constant ε such that

$\|x_i(n)\| < \varepsilon$ for any n and $\|x_b(k)\| < \varepsilon$ for all $k \ge n$

Now, consider the boundedness of the zero-input response. Taking a norm of that response and using the fact (from equation (2.8)) that

$$\Phi_i(k, n) = \prod_{m=n}^{k-1} \sigma_i(m)$$

the following inequalities are obtained:

$$\|\Phi_i(k, n)x_i(n)\| \le |\Phi_i(k, n)| \cdot \|x_i(n)\|$$

$$< \varepsilon \prod_{m=n}^{k-1} |\sigma_i(m)|$$

Since $|\sigma_i(k)| \le 1$ for all $k \ge n$, there exist some constants ρ ($0 < \rho \le 1$) and r ($0 < r < 1$) such that

$$0 \le \prod_{m=n}^{k-1} |\sigma_i(m)| \le \rho r^{k-n} \le \rho$$

Hence,

$$\|\Phi_i(k, n)x_i(n)\| < \varepsilon\rho$$

Next consider the boundedness of the zero-state response. Taking a norm of that response and using the fact that $0 \le \xi_i(k) \le 2$ for all $k \ge n$ yields

$$\left\|\sum_{j=n+1}^{k}\Phi_i(k,j)\xi_i(j-1)\mathbf{x}_b(j-1)\right\| \le \sum_{j=n+1}^{k}|\Phi_i(k,j)|\cdot|\xi_i(j-1)|\cdot\|\mathbf{x}_b(j-1)\|$$

$$< \varepsilon\sum_{j=n+1}^{k}\prod_{m=j}^{k-1}|\sigma_i(m)|\cdot|\xi_i(j-1)|$$

$$\le 2\varepsilon\sum_{j=n+1}^{k}\rho r^{k-j} < 2\rho\varepsilon\lim_{k\to\infty}\sum_{j=n+1}^{k}r^{k-j} < \frac{2\rho\varepsilon}{1-r}$$

Hence, combining the above results for two responses shows that the reproduction operator is stable.

As an example of the new reproduction operation, consider the problem of maximising the following fitness function $f(x_1, x_2)= 4.5 - x_1^2 - x_2^2 \ge 0$, subject to $0 \le x_1, x_2 \le 1.5$. x_1 and x_2 are encoded as 4-bit binary strings and then concatenated to form a structure. x_1 and x_2 have a resolution of 0.1. One possible reproduction operation on a population of these structures with $\eta_i= 1$ ($1 \le i \le 6$) is illustrated in Table 2.1. The 5th structure shown in Table 2.1 is the best and is directly transferred into the mating pool. The other structures are reproduced using equation (2.5).

Table 2.1 One possible reproduction operation of the HGA

No.	Current population P(k)			New population $\overline{\mathbf{P}}$(k+1)	
i	$s_i(k)$	$x_i(k)$	$f_i(k)$	$\overline{x}_i(k+1)$	$\overline{s}_i(k+1)$
1	00111001	0.3 , 0.9	3.60	0.3 , 0.8	00111000
2	10100110	1.0 , 0.6	3.14	0.9 , 0.5	10010101
3	00101001	0.2 , 0.9	3.65	0.2 , 0.8	00101000
4	10011110	0.9 , 1.4	1.73	0.7 , 0.8	01111000
5	**01100011**	**0.6 , 0.3**	**4.05**	**0.6 , 0.3**	**01100011**
6	11100010	1.4 , 0.2	2.50	1.1 , 0.2	10110010

Like other reproduction operators, this reproduction operator must also be used with some care. When the overall performance becomes better and individuals cluster to a point, the GA may encounter difficulty in differentiating between good individuals and better ones. This retards the rate of convergence towards the solution. In this case, fitness scaling is necessary to enhance the solution efficiency. For this purpose, the scaling window scheme [Grefenstette, 1986] is adopted.

Crossover and Mutation Operators. Once structures have been produced to form the mating pool, new search points are further explored by two subsequent operators, i.e., crossover and mutation. The two operators adopted for the HGA implementation are the simple one-point crossover operator and the simple mutation operator.

Procedure of the Hybrid GA. The overall procedure of the Hybrid GA is as follows:

HYBRID_GENETIC_ALGORITHM

Set k= 0;
Create an initial population $P(k)$ of size N randomly;
Decode individuals $s_i(k)$ $(1 \leq i < N)$ into parameter vectors $x_i(k)$;
Evaluate fitness $f_i(k)$ $(1 \leq i \leq N)$ incorporating the scaling window scheme;
Determine the best $x_b(k)$ providing the largest fitness $f_b(k) = \max_{1 \leq i \leq N} [f_i(k)]$;

WHILE < the termination conditions are not met >
Assign a new vector $\bar{x}_i(k+1)$ to each individual using the formula

$$\bar{x}_{ij}(k+1) = x_{ij}(k) + \eta_i [(f_b(k) - f_i(k)) / f_b(k)] [x_{bj}(k) - x_{ij}(k)] \quad (1 \leq i \leq N, 1 \leq j \leq M)$$

where $\bar{x}_{ij}(k+1)$, $x_{ij}(k)$ and $x_{bj}(k)$, respectively, are the jth elements of

$\bar{x}_i(k+1)$, $x_i(k)$ and $x_b(k)$, and η_i is the ith positive coefficient;
Encode $\bar{x}_i(k+1)$ $(1 \leq i \leq N)$ into structures $\bar{s}_i(k+1)$ to form a pool $\overline{P}(k+1)$;

Crossover $\overline{P}(k+1)$ to form the intermediate population $\tilde{P}(k+1)$;

Mutate $\tilde{P}(k+1)$ to form the new population $P(k+1)$;
Decode individuals $s_i(k+1)$ $(1 \leq i \leq N)$ into parameter vectors $x_i(k+1)$;
Evaluate fitness $f_i(k+1)$ $(1 \leq i \leq N)$ incorporating the scaling window scheme and determine the best $x_b(k+1)$ (at this step, output the potential solution);
Set k= k+1;
END WHILE

Normally, the iterations proceed until the solution is encountered with specified accuracy or a predefined number of iterations is reached.

2.1.2 Cross-Breeding in Genetic Optimisation

A problem encountered in applying GAs is premature convergence to non-optimal solutions. A method of avoiding this problem is to explore several different areas of the solution space simultaneously by evolving different sub-populations of solutions in parallel. The degree of interaction between the sub-populations is important to the performance obtained, as too much interaction can reduce the benefit of using different sub-populations in that the situation can revert to that of having a single large population.

This section presents a simple GA which allows limited interaction between the sub-populations through a one-step "cross-breeding" procedure where selected members of the sub-populations are intermingled to form a combined population for further optimisation. The algorithm avoids disturbing the evolution process of each sub-population by allowing it to "mature" sufficiently before carrying out cross-breeding. The proposed algorithm thus achieves better results than ordinary GAs without significantly increasing computational overheads.

In this section, previous work in the related area of parallel genetic algorithms (PGAs) is overviewed. Then a cross-breeding GA is described.

Related Work on Parallel Genetic Algorithms. There are two methods of obtaining a parallel genetic algorithm (PGA). The computational tasks in an ordinary GA can be divided up and distributed to several processors [Maruyama *et al.*, 1991; Baluja, 1993]. The result is a faster GA with otherwise the same performance as the original GA. Alternatively, a number of processors can be employed, each attempting to optimise a sub-population and communicating its outputs to the other processors [Muhlenbein, 1989]. This method can have the desired effect of avoiding premature convergence to non-optimal solutions as it involves simultaneously searching different areas of the solution space. In an extensively parallel GA, both methods can be combined.

The size of the sub-populations used determines whether a PGA is fine-grained or coarse-grained. Some PGAs operate with individual solutions in parallel (fine-grained PGAs [Maruyama *et al.*, 1991; Baluja, 1993, Manderick and Spiessens, 1989]). That is, they have sub-populations with sizes of 1. Other PGAs, the coarse-grained PGAs, have sub-populations of sizes larger than 1 [Baluja, 1993]. In general, there are more parallel processes and more interactions in a fine-grained PGA. To limit interactions, some fine-grained PGAs adopt a localisation approach where individual solutions are assigned to points on a 2D-grid and are only allowed to communicate (or mate) with their neighbours on the grid [Maruyama *et al.*, 1991]. This approach is also employed in some coarse-grained PGAs where a processor looking after a sub-population only copies its best individual solutions to neighbouring processors [Muhlenbein, 1989; Starkweather *et al.*, 1991].

A study of PGAs has found that, for a number of optimisation problems, they yielded better solutions than the ordinary GA [Gordan and Whitely, 1993], although the main thrust of the parallelisation was to increase speed rather than produce improved solutions. A disadvantage with the schemes reviewed in [Baluja, 1993; Muhlenbein, 1989; Starkweather *et al.*, 1991] is their complex interaction procedure which could hamper the search for the global optimum.

A Cross-Breeding Genetic Algorithm. The flowchart of a cross-breeding GA is depicted in Figure 2.4. Blocks 1 to n are each identical to the flowchart shown in Figure 1.12 in Chapter 1. These blocks represent n GAs executing independently of one another. The initial sub-populations for these GAs are created using random number generators with different "seeds". After a given number of generations, MAXGEN1, the execution of these GAs is stopped. MAXGEN1 is normally chosen to be sufficiently large to allow the individual sub-populations to "mature". The n populations of solution strings produced by these GAs represent n sets of "preliminary" designs. A "cross-breeding" procedure is applied to create a new population from these preliminary designs. Various methods could be used for this purpose, including a "greedy" method where the best solutions from each sub-population are always selected for the new population and a method based on the "biased roulette wheel" technique where the better solutions are only given greater chances of being selected.

2.1.3 Genetic Algorithm with the Ability to Increase the Number of Alternative Solutions

The main advantage of genetic algorithms is that they can usually converge to the optimal result even in complex multimodal optimisation problems. For such problems, a standard GA tends to cluster all of its solutions around one of the peaks, even if all peaks have identical heights.

There are situations where it is necessary to locate more than one peak because different alternative solutions are wanted. This section presents four techniques to help a GA find multiple peaks. These are: using a "sharing" mechanism to achieve a more even distribution of 'chromosomes' around the potential peaks, employing a "deflation" mechanism to penalise certain solution chromosomes within the chromosome population to further increase genetic variety, eliminating identical chromosome strings, and applying heuristics to encourage convergence to good solutions.

"Sharing" Mechanism. The basic idea of a "sharing" mechanism is to force individuals located close to one another to share the "local" available resources until an equilibrium among different areas is reached [Goldberg and Richardson, 1987]. Ideally, this should result in a distribution of individuals which is proportional to the amount of local resources available, in other words an even distribution of solutions around several peaks at once.

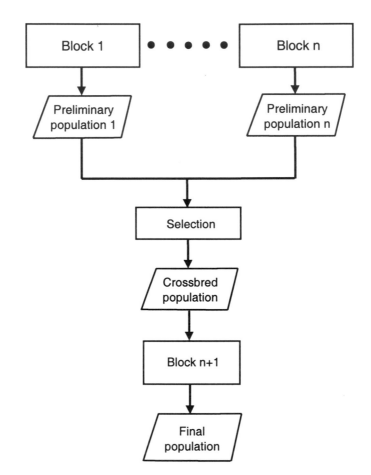

Fig. 2.4 Flow chart of the cross-breeding GA

If F_a and F_b are the amounts of resources at locations A and B respectively, a simple "sharing" mechanism gives the equilibrium criterion as:

$$\frac{F_a}{N_a} = \frac{F_b}{N_b}$$

(2.9)

where N_a and N_b are the numbers of individuals queuing up at locations A and B respectively. For example, if $F_a = 1000$, $F_b = 700$ and $N_a = 10$, then the equilibrium point can only be reached when $N_b = 7$.

To implement "sharing", a metric for measuring the distance between two individuals is required. It is easy to devise such a metric for a single-variable continuous function, for example the function $F(x) = \sin^6(5.1\pi x + 0.5)$ used in [Goldberg and Richardson, 1987]. In that case, the absolute value of the difference between the x values can be adopted as the distance metric d_{ij} (the phenotype distance [Goldberg and Richardson, 1987]), that is:

$$d_{ij} = d(x_i, x_j) = \left| x_i - x_j \right|$$

(2.10)

where x_i and x_j are decoded parameters which themselves are functions of the strings S_i and S_j.

The sharing function is defined in [Goldberg and Richardson, 1987] as:

$$sh(d_{ij}) = \begin{cases} 1 - (d_{ij}/\sigma_s)^\alpha & d_{ij} < \sigma_s \\ 0 & \text{otherwise} \end{cases}$$

(2.11)

where σ_s and α are constants; σ_s defines the zone in which the sharing function is non-zero and α controls the shape of the sharing function, that is how quickly it reduces to zero. Both σ_s and α are problem specific and again there is some freedom in choosing their values.

The above sharing function has the following three properties:

1. $0 \le sh(d_{ij}) \le 1$, *for all* d_{ij}

2. $sh(0) = 1$ (2.12)

3. $\lim\limits_{d_{ij} \to \alpha_s} sh(d_{ij}) = 0$

The shared fitness $\hat{F}_i$ of a string i is its potential fitness F_i divided by its "niche count" $\hat{m}_i$:

$$\overset{\wedge}{F_i} = \frac{F_i}{\overset{\wedge}{m_i}} \tag{2.13}$$

The niche count $\overset{\wedge}{m_i}$ for a particular string i is set to the sum of all sharing function values taking the entire population of N potential solutions:

$$\overset{\wedge}{m_i} = \sum_{j=1}^{N} sh(d_{ij}) \tag{2.14}$$

When there is no sharing, $\overset{\wedge}{m_i} = 1$, and the string is awarded the full potential fitness value.

Otherwise, $\overset{\wedge}{m_i}$ is larger than 1 and the potential fitness value of the string is to be shared according to the closeness and the number of its neighbouring points [Goldberg and Richardson, 1987].

The above method of resource sharing by dividing fitness values among individual strings ("division-based sharing") is strictly applicable only when these values are proportionally related to the quality of the strings. Where this is not the case, as in the gearbox example described later in Section 2.2, the spirit of sharing can still be implemented by appropriately reducing the potential fitnesses of the strings among which sharing takes place. A possible reduction strategy ("reduction based sharing") is to deduct an amount dependent on a string's niche count from its potential fitness, viz.

$$F_i^\alpha = F_i - W(\overset{\wedge}{m_i} - 1) \tag{2.15}$$

where F_i^α is the adjusted fitness value of the ith string, F_i the potential fitness value, W a weighting factor and $\overset{\wedge}{m_i}$ the niche count. Clearly, when there is no crowding of strings around one location, $\overset{\wedge}{m_i} = 1$ and the individual is awarded its full potential fitness value. It is necessary to point out the implicit use of the continuity property of the fitness function in applying the phenotype distance metric, as done in the example described above. For a continuous function, $y = F(x)$,

$$\lim \Delta y = 0 \quad \text{when } \Delta x \to 0$$

so that two individuals x_i and x_j located close to each other have similar fitness values y_i and y_j. Thus, Δx could be used to measure the distance between individuals for the purpose of resource sharing.

Compared with the case of single-variable continuous functions, "sharing" for multiple-variable discrete functions is more difficult to implement. The major difficulty is the definition of the distance, d_{ij}, between two individuals, i and j, each of which is characterised by a set of parameters instead of a single variable. Due to the discontinuity of the fitness function, the phenotype distance metric will not reflect the distance between the locations of the resources available for a pair of individuals.

How the distance metric is designed will depend on the application. An example of a distance metric for a multi-variable discrete fitness function will be given in Section 2.2 when the gearbox design problem is discussed.

"Deflation" Mechanism. In algebra, one way of finding the set of roots (R) for a non-linear equation, $F(x) = 0$, is to find one root x_0 first and then eliminate that root from the equation by division:

$$\hat{F}(x) = \frac{F(x)}{x - x_0} \tag{2.16}$$

This operation, known as *deflation* [Brown and Gearhart, 1971], reduces the problem to finding the set of roots ($\hat{R}$) for $\hat{F}(x) = 0$. Note that x_0 has been eliminated from $\hat{R}$ and $R = \hat{R} \cup \{x_0\}$.

The deflation method in a GA works in a similar manner. A fixed amount D is artificially deducted from the fitness value of a satisfactory individual which has been marked and recorded. This reduces that individual's chance of being reproduced. It is in this sense that such an individual is said to have been "eliminated". Note that the "elimination" process does not alter the structure of the individuals being eliminated. Reduction of the fitness value by a fixed amount maintains the relative ranking among the solutions being "eliminated".

This measure has two desirable effects. First, it prevents the domination of one superior individual during the evolution process. Second, it gives rise to a greater variety of solutions.

Although those "eliminated" individuals are no longer regarded as satisfactory solutions due to their artificially reduced fitness values, they are actually still "genetically" good strings. It would be advantageous to keep them in the reproduction cycle even after the deflation procedure, so that their merits can be utilised and combined with those of newly emerging individuals to give new and stronger individuals. To achieve this, the amount of deflation, D, should be just enough to enable the not-so-fit solutions to emerge into the reproduction cycle.

Identical String Elimination. Although identical individuals are rare in a randomly generated initial population, duplication of individuals will happen during the evolution process. This is due to the nature of the reproduction process in the GA. With a fixed population size, identical individuals reduce the diversity of the population and, hence, bias the reproduction process to a certain location and possibly lead to premature convergence. It would also be meaningless to apply sharing among identical individuals. In order to obtain more variety, a method called "identical string elimination" is employed. With this technique, a heavy reduction in fitness value is imposed on each group of identical individuals, leaving only one representative of the group unaffected. Keeping one representative is necessary for passing on useful "genetic" information to the population.

The reduction method here is different from that used in deflation. Instead of a fixed reduction, a "proportional reduction" method is employed. The reduction performed is a percentage (I) of the potential fitness value of the individual. The reason for selecting this method is quite simple. Given the wide span of fitness value levels at which groups of identical individuals may occur, it is impossible to find a fixed reduction amount. A severe reduction can effectively "eliminate" the stronger individuals. At the same time, it may also cause the weaker individuals to have negative fitness values. This conflicts with the convention that fitness values should be positive [Mansfield, 1990]. Moreover, if it is decided to make all negative fitness values equal to zero, there would be no distinction between weak individuals with fitnesses below a certain level.

Heuristics. It has been reported elsewhere [Powell *et al.*, 1990] that GAs may be inefficient or even unable to converge with a purely randomised initial population. This situation can be improved by introducing heuristics into the GA to form a hybrid search system. Heuristics can help the GA in two ways. First, they can position the initial population in a near-optimum area so that the algorithm can reach the optimum quickly. Second, by giving some direct guidance to the search they can also substantially reduce the searching time.

These heuristics are problem dependent. For example, in the gearbox design problem, heuristics have been devised that correct strings that have low fitness values due to one of the design parameters being inappropriate. A typical heuristic is:

IF (TotalRatio - 700.0/OutputSpeed) < 1.0 AND (InputSpeed ≠ 700.0)
THEN Make InputSpeed = 700.0

Application of this heuristic causes the input speed parameter of the string to be changed to give the correct input speed (700 r/min in this case) and thus dramatically increase the fitness of the string.

Unfortunately, heuristics can also bias the performance of a GA; that is, in a hybrid GA implementation, heuristics serving as an external force can accelerate to a certain location. In order to reduce this undesirable side effect of the heuristics, two countermeasures are available. The first is to apply the heuristics probabilistically according to a predetermined rate, similar to the crossover rate or mutation rate, that is to reduce the frequency of applying heuristics. The second is to delay the application of heuristics by a number of generations and thus allow the formation of a relatively even population distribution. The method is based upon the assumption that a heuristic-free GA would conduct a smoother search than a hybrid algorithm.

The flow chart of a genetic algorithm implementing the techniques described above is presented in Figure 2.5.

2.1.4 Genetic Algorithms with Variable Mutation Rates

There are five factors influencing the performance of a GA: the method of representing solutions (how solutions are encoded as chromosomes), initial population of solutions (the group of chromosomes created at the beginning of the evolution process), evaluation function (the metric for measuring the fitness of a chromosome), genetic operators (for example, the three common operators, selection, crossover and mutation, for choosing chromosomes for a new generation, creating new chromosomes by combining existing chromosomes, and producing a new chromosome by altering an existing chromosome), and control parameters (for example, the size of the population of chromosomes, and rates of crossover and mutation).

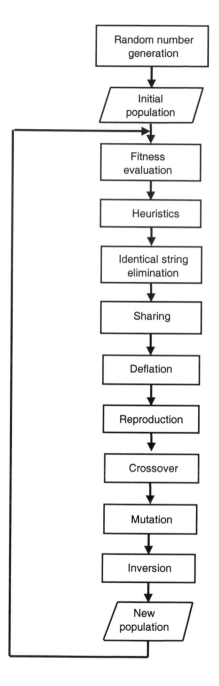

Fig. 2.5 Flow chart of a genetic algorithm able to find multiple peaks

The first three factors and, to some extent, the fourth factor also, are problem dependent. For instance, in a problem where there is prior knowledge of reasonably fit solutions, they can be used to form the initial population to speed up the optimisation process. In other problems, such as the Travelling Salesman Problem, where chromosomes cannot contain duplicated genes (for example, the sequence of cities to be visited by the salesman cannot include a given city more than once), some genetic operators, like the crossover operator, are not usable.

The fifth factor, GA control parameters, tends to be much less problem dependent and there is more scope for generic work aimed at improving the performance of a GA by manipulating its control parameters. This section focuses on one control parameter in particular, the mutation rate. Mutation is an operator used to create diversity in the population of solutions. It is a search operator for exploring the solution space during optimisation. A correct mutation rate is required, as too high a rate converts the GA into a random search procedure and too low a rate can cause the process to be trapped at local optima.

In this section, previous work on control parameters is first summarised. Then three new strategies for varying the mutation rate are described.

Previous work on control parameters. De Jong [1975] has carried out experiments on different test problems to study the effect of control parameters on the performance of a GA. He has suggested a set of parameter values (see Table 2.2) which have been found generally to produce good on-line performance (the average performance of all solutions obtained in all generations) and off-line performance (the performance computed by using only the current best solution value for each generation).

Similarly, Schaffer et al. [1989] have conducted a wide-ranging experiment on the effect of control parameters on the performance of a GA and concluded that good on-line performance can be obtained using the set of parameters given in Table 2.2.

Grefenstette [1986] has investigated the use of a GA to optimise the control parameters for another GA. This method of determining control parameter values is considered more robust than the experimental approach of De Jong and Schaffer et al. as it better explores the space of parameter values. The set of parameter values found by Grefenstette is presented in Table 2.2.

A theoretical investigation of the optimal population size for genetic algorithms employing binary-coded chromosomes has been carried out by Goldberg [Goldberg, 1985] who found the following relation between population size and length of a chromosome :

Population size $= 1.65 \times 2^{0.2 \text{ length}}$

Table 2.2 Suggested values for control parameters of genetic algorithms

Control parameters	De Jong	Schaffer	Grefenstette
Population size	50-100	20-30	30
Crossover rate	0.60	0.75-0.95	0.95
Mutation rate	0.001	0.005-0.01	0.01

The above studies have focused on determining good control parameter values for genetic operators. Once they have been found, they remain fixed for all applications. However, the optimal parameter setting is likely to vary for different problems. Therefore, to achieve a good performance, new optimal sets of values should be obtained for a given problem, which is a time consuming task. Employing parameters with automatically variable values can save time and yield improved performance. There have been a number of studies on adapting the control parameters during optimisation. Davis [1989] has described a technique for setting the probabilities of applying genetic operators, and therefore their rates, while the GA is executing. The technique involves adjusting the probabilities for the different operators based on their observed performance as the optimisation proceeds. Davis's technique is complex as it does not just involve the mutation operator but other operators also.

Whitely and Hanson [1989] have proposed another form of adaptive mutation. This involves indirectly monitoring the homogeneity of the solution population by measuring the Hamming distance between parent chromosomes during reproduction: the more similar the parents, the higher the probability of applying the operator to their offspring so as to increase diversity in the population. However, as the population naturally contains more multiple copies of good chromosomes towards the end of evolution, this strategy can be disruptive at that stage and therefore delay convergence.

Fogarty [1989] has studied two variable mutation regimes. The first is to use different mutation probabilities for different parts of a chromosome (higher probabilities for less significant parts). However, the probabilities are fixed for all generations and all chromosomes. In the second regime, the same mutation probability is adopted for all parts of a chromosome and then decreased to a constant level after a given number of generations. Both of Fogarty's strategies are non-adaptive in the sense that mutation rates are all pre-determined. Thus, these strategies are implicitly based on the assumption that the evolution process always follows a fixed pattern, which is clearly not true.

Strategies for Varying Mutation Rates. Three new heuristic strategies are described in this section. The first strategy involves changing the mutation rate when a "stalemate" situation is reached. The second strategy adapts the mutation rate according to the fitness of a chromosome. The third strategy embodies the second strategy but also prescribes different mutation probabilities for different parts of a chromosome.

First Strategy: With this strategy, the minimum ($mutrate_{min}$) and maximum ($mutrate_{max}$) mutation rates are first decided by the user who also selects a threshold ($nrep_{max1}$) for the number of generations $nrep$ during which the performance of the best solution can remain the same before the mutation rate has to increase. At the same time, the user chooses a threshold ($nrep_{max2}$) for the number of generations required to ramp up to the maximum mutation rate.

At the beginning of the optimisation, the mutation rate $mutrate$ is set to $mutrate_{min}$. As the optimisation proceeds, the performance of the best solution in each generation is recorded and if it has not improved after $nrep_{max1}$ generations (that is, a "stalemate" situation is deemed to have occurred), the mutation rate is increased gradually according to the following equation:

$$mutrate = \frac{nrep - nrep_{max1}}{nrep_{max2}} (mutrate_{max} - mutrate_{min}) + mutrate_{min} \qquad (2.17a)$$

where $(nrep - nrep_{max1}) \leq nrep_{max2}$

The purpose of increasing the mutation rate is to enhance the probability of finding new improved solutions. As soon as an improvement in the performance of the best solution is observed, the mutation rate is directly reduced back to the $mutrate_{min}$ value, which is maintained until a new "stalemate" situation is reached. This strategy is illustrated in Figure 2.6(a). The figure also shows that the mutation rate is reduced to $mutrate_{min}$ from $mutrate_{max}$ even though an improved solution has not been found. This is to avoid excessive disruption to the evolution process.

In the initial stages of an optimisation process, say the first 50 generations, it is likely that there are few good solutions in the population. Therefore, it would be preferable to employ a high mutation rate to accelerate the search. Thus, a better strategy is depicted in Figure 2.6(b) which involves starting with the maximum mutation rate $mutrate_{max}$ and reducing it gradually to $mutrate_{min}$. The mutation rate used during a generation g in this initial phase is given by the following equation :

$$mutrate(g) = (mutrate_{max} - c.g) + mutrate_{min} \qquad (2.17b)$$

where c is a constant and g is the current generation number, which should be less than or equal to ($mutrate_{max}/c$).

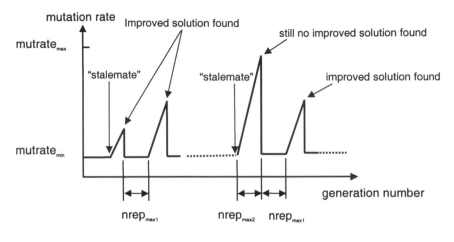

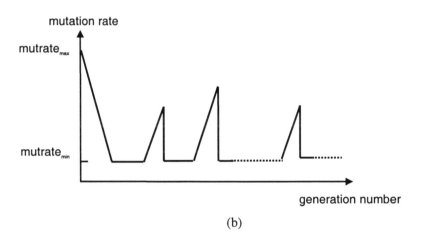

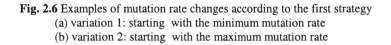

(a)

(b)

Fig. 2.6 Examples of mutation rate changes according to the first strategy
(a) variation 1: starting with the minimum mutation rate
(b) variation 2: starting with the maximum mutation rate

The mutation rate value computed using equation (2.17a) or (2.17b) is applied to all chromosomes in a given generation.

Both variations of the first strategy should help the GA to escape from local optima and find improved solutions in new areas of the solution space. When a promising solution is found, the mutation rate is reduced to enable the GA to search those areas in more detail.

Second Strategy: This strategy is based on the idea that poor solutions require more correction than better solutions. Thus, the fitness value *fit(i)* of each chromosome *i* can be used to determine the mutation rate associated with it. As with the first strategy, an upper limit and a lower limit are chosen for the mutation rate and within those limits the mutation probability *mutrate(i)* for chromosome *i* is calculated according to its fitness value: the higher the fitness value, the lower the mutation rate. A suitable equation for calculating *mutrate(i)* is :

$$mutrate(i) = (1\text{-}fit(i))\ (mutrate_{max} - mutrate_{min}) + mutrate_{min} \qquad (2.18)$$

In equation (2.18), the fitness value *fit(i)* is assumed to have been normalised in the range 0.0 to 1.0. This strategy is shown in Figure 2.7. When using this strategy, the mutation operation is carried out before other operations that can change the chromosomes in order to reduce the number of times fitness values have to be calculated.

Third Strategy: This strategy is applicable in optimisation problems where the genes of a chromosome represent individual numerical parameters. As with the second strategy, this strategy also involves mutating chromosomes according to their fitness. However, instead of assigning all parts of a gene the same mutation probability, the left-most digits which are the most significant are initially given a greater chance of being mutated. This is because, at the beginning of the optimisation, promising areas of the solution space have not yet been located and the solutions are still poor and need more correction. When a good solution is found, the mutation rates for the left-most digits are reduced compared to those for the right-most digits as the solution does not require so much correction. This leads to the fine tuning of solutions.

Thus, with this strategy as illustrated in Figure 2.8, the mutation probability for each bit in a chromosome depends on three parameters: the fitness of the chromosome, the generation number and the position of the bit along the particular gene of the chromosome. Note that the second and third parameters are the same as those employed in Fogarty's second and first strategies respectively. However, the use of the fitness of a chromosome to determine its mutation rate makes the proposed strategy adaptive, unlike Fogarty's strategies, as mentioned previously.

Assuming a binary representation is adopted for the chromosomes, using this strategy, the mutation rate for bit k ($k = 1$ to l_g) of chromosome i is computed as follows:

$mutrate(i,k) = (k/lg)\, d$ if $g \leq maxgen/2$ (2.19a)

$mutrate(i,k) = (1/k)\, d$ if $g > maxgen/2$ (2.19b)

where $d = (1\text{-}fit(i))\,(mutrate_{max} - mutrate_{min}) + mutrate_{min}$. l_g is the gene length, *maxgen* is the maximum number of optimisation generations to be tried and g is the current generation number.

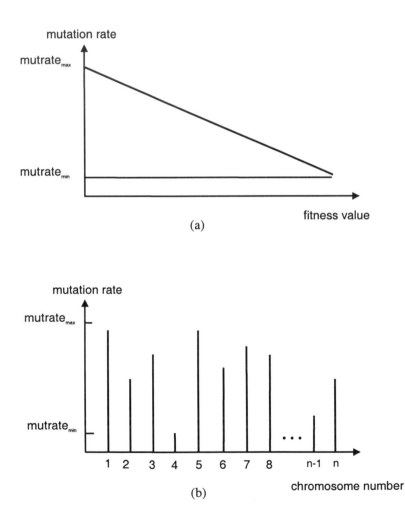

(a)

(b)

Fig. 2.7 Illustration of the second strategy: (a) mutation rate as a decreasing function of fitness value, (b) examples of mutation rates for a generation of chromosomes

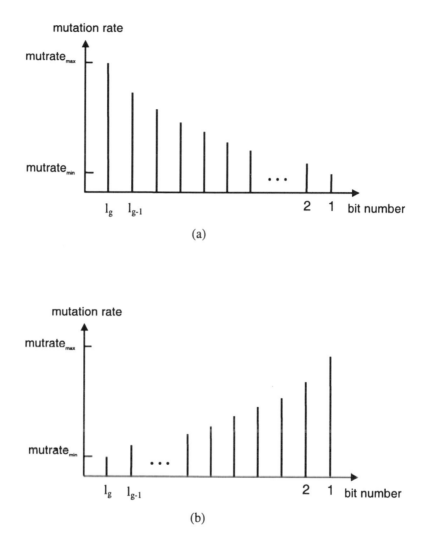

Fig. 2.8 Third strategy: mutation rates for a chromosome (a) before and (b) after the half-way point (generation maxgen/2) of the evolution process

2.2 Engineering Applications

This section describes applications of genetic algorithms to problems from different engineering areas. These include static and adaptive fuzzy logic controller design [Pham and Karaboga, 1991a; Pham and Karaboga, 1994b], neural network training for dynamic system identification [Pham and Karaboga, 1994a], gearbox design [Pham and Yang, 1993] and automatic generation of workplace layouts [Pham *et al.*, 1996].

2.2.1 Design of Static Fuzzy Logic Controllers

This section describes the use of genetic algorithms (GAs) in the design of fuzzy logic controllers (FLCs). First the way in which the problem of designing an FLC can be treated as an optimisation problem is explained. Second the procedure for designing a Proportional FLC based on a standard GA is detailed and the results obtained using the procedure are given. Third, a Proportional-plus-Integral (PI) type of FLC is also designed and the simulation results obtained using this controller are presented.

Fuzzy Logic Controller Design as an Optimisation Problem. A simple FLC is affected by several factors [Lee, 1990a-b]. These include the input fuzzy membership functions, the relation matrix representing the rule base, the inference mechanism, the output quantisation levels, the input and output mapping coefficients and the defuzzification method adopted. In order to design an efficient FLC for a process, these factors should be selected carefully. Usually, to simplify the design problem, some of the factors are pre-determined and only a subset are variable and selectable. For more information about fuzzy controllers, see Appendix 2.

Let q be a vector of all the variable parameters of an FLC. The space Q of possible vectors q is potentially very large. The design of the FLC can be regarded as the selection of an optimal vector $\hat{q}$ that will minimise a cost function $J(T_C)$ subject to the constraint that $\hat{q}$ belongs to Q, viz.

$$\underset{q \in Q}{Min}\, J(T_C) \tag{2.20}$$

An example of a simple cost function might be:

$$J(T_C) = \sum_{t=1}^{T_C} (|\, y_d(t) - y_p(t)\,|) \tag{2.21}$$

where y_d and y_p are the desired and actual outputs of the plant and T_C is the total time over which cost is computed.

In general, the constrained design space Q is a complex multidimensional space and the function $J(T_C)$ is non-linear and multimodal. To find the optimal vector $\hat{q}$ requires a robust search algorithm capable of efficiently exploring such a space.

Applying a Genetic Algorithm to the Problem. There are many versions of GAs [Krishnakumar, 1989; Davis, 1991; Goldberg, 1989; Muselli, 1992; Pham and Karaboga, 1991b] and, in the following, both the optimisation problem and the GA used will be explained. To show the performance of the new design technique, comparative results will be presented for a time-delayed second-order process controlled by an FLC designed using a GA, an FLC designed using Peng's technique [Peng, 1990] and a conventional PID controller.

Optimisation Problem: The key element in an FLC is the relation matrix which represents the original set of rules [Pham and Karaboga, 1991a]. This matrix defines a mapping from the space of errors between the desired and actual process outputs to the space of control actions. The performance of the controller thus hinges on the relation matrix which, in turn, depends on the rules from which it was constructed. Consequently, one of the main problems in the design of FLCs is how to obtain a proper relation matrix. Depending on the process to be controlled, the relation represented by the matrix could be simple or complex and non-linear.

The dimensions of the relation matrix are related to the cardinality of the input and output universes (or spaces) or, if the latter are continuous universes, the number of quantisation levels used. The number of rows of the matrix is described by the nature of the input universe and the number of columns by that of the output universe. Relation matrices for many FLCs reported in the literature have dimensions equal to 7×11. Each element of a relation matrix is a fuzzy membership value and thus is a real number in the range 0 to 1. In this work, relation matrices of dimensions 7×11 were employed and eight bits were used to code an element of a relation matrix. Therefore, a relation matrix was represented as a binary string of 8×7×11 or 616 bits. Figure 2.9 shows an example of a binary representation of a relation matrix.

Genetic Algorithm Used: The GA version adopted here is that defined by Grefenstette [Grefenstette, 1986]. The flowchart of the algorithm is given in Figure 1.12.

Each individual in the population of solutions is sent to the evaluation unit and its fitness is calculated to determine how well it would perform as a controller for the given process.

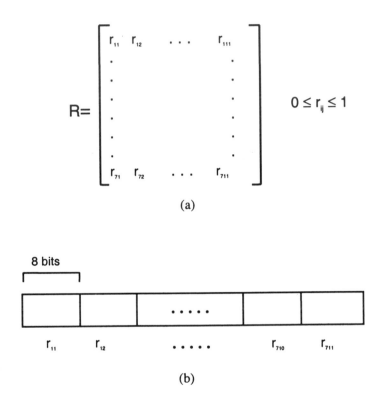

(a)

(b)

Fig. 2.9 (a) A fuzzy relation matrix R
(b) String representation of R

To do this, the relation matrix corresponding to the individual string is reconstructed and used to control the process (or a model of it). The performance of the system is then computed using the *ITAE* (*Integral of Time Multiplied by Absolute Error*) criterion, viz.:

$$ITAE = \sum_{i=1}^{L} i\,T\,\left|y_d[i] - y_p[i]\right|$$ (2.22)

where y_d is the desired output of the process, y_p is the actual output of the process under control, L is the simulation length and T is the sampling period.

The fitness f of the individual is then evaluated from *ITAE* as:

$$f = (ITAE + C_c)^{-1}$$ (2.23)

where C_c is a small constant (0.01) added to *ITAE* as a precautionary measure to avoid dividing by zero.

Description of FLC Designed: The input space of errors of the controller was divided into seven fuzzy partitions, each of which was represented by a fuzzy value: Negative Big (NB), Negative Medium (NM), Negative Small (NS), Zero (ZE), Positive Small (PS), Positive Medium (PM) and Positive Big (PB). The output space was quantised into eleven different levels. Both spaces were assumed to span the interval [-5,+5]. The membership functions used are given in Figure 2.10.

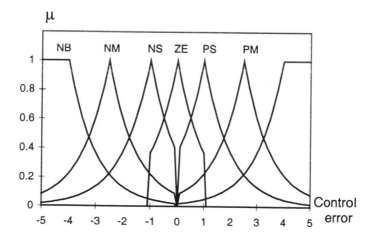

Fig 2.10 Membership functions used

The inference mechanism employed the Max-Product rule. Defuzzification was implemented using the Centre of Area method.

Results: Fuzzy logic controllers were to be designed for a time-delayed second-order plant with the following transfer function:

$$G(s) = (K\,e^{-\tau s})/(as+b)^2 \tag{2.24}$$

It was assumed that a sampled data system with a zero-order hold (ZOH) was adopted. The sampling period, T, was selected as 0.1 seconds. The plant characteristic parameters were:

$$\tau = 0.4, \quad a = 0.3, \quad b = 1.0, \quad K = 1.0 \tag{2.25}$$

Figure 2.11 shows the simulated step response of the plant defined by equations (2.24) and (2.25) when under the control of an FLC designed by the GA.

For comparison, Figure 2.12 shows the response obtained from the same plant using a conventional PID controller with the following parameters:

$$K_p = 0.630517, \quad T_i = 0.594813, \quad T_d = 0.237036 \tag{2.26}$$

These parameters are *"optimal"* parameters for the given plant [Peng *et al.*, 1988]. The sampling period was 0.02 seconds for the PID control system.

The results achieved for this plant using a fuzzy PID controller optimised using the parametric function optimisation method [Peng, 1990] are:

Rise time = 2.8 seconds
Overshoot = 4% $\tag{2.27}$
Settling time for $\pm$ *3 % error = 5 seconds*

Figures 2.13 and 2.14 show the performance of the PID and FLC controllers when uniform white noise with variance equal to 0.0132 was applied to the output signal of the control system.

The responses of both the PID and FLC control systems in the case of process parameter changes are presented in Figures 2.15(a-d) and 2.16(a-d). The responses in Figures 2.16(a-d) were obtained for an FLC designed to control the process defined by equations (2.24) and (2.25).

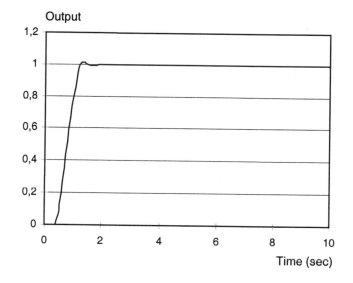

Fig. 2.11 Response of plant using an FLC designed by the standard GA

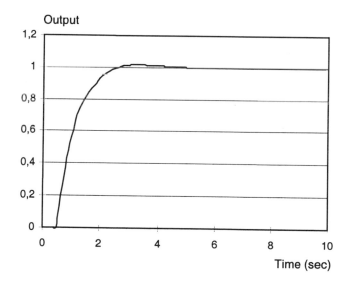

Fig. 2.12 Response of the plant using a PID controller

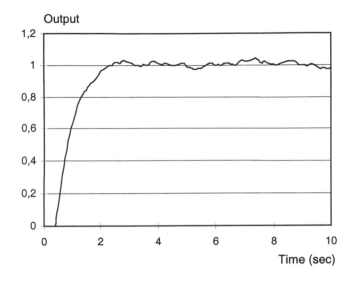

Fig. 2.13 Response of plant using PID controller (white noise present)

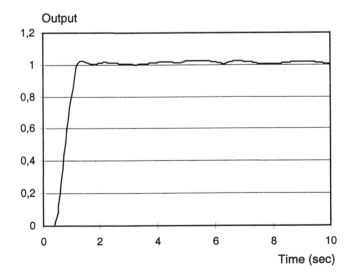

Fig. 2.14 Response of plant using FLC (white noise present)

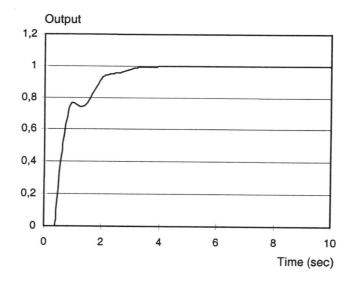

Fig. 2.15 (a) Response of plant using PID controller (a=0.4; other process parameters as given by equation (2.25))

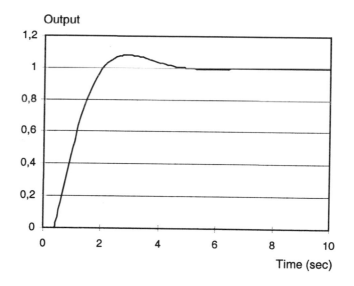

Fig. 2.15 (b) Response of plant using PID controller (a=0.2)

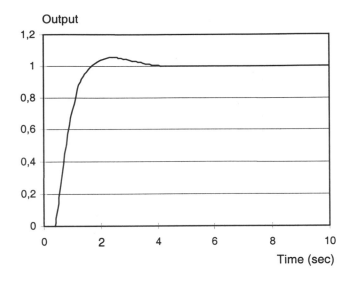

Fig. 2.15 (c) Response of plant using PID controller (b=0.9)

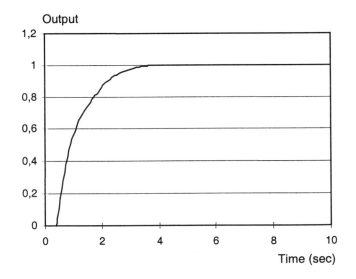

Fig. 2.15 (d) Response of plant using PID controller (b=1.1)

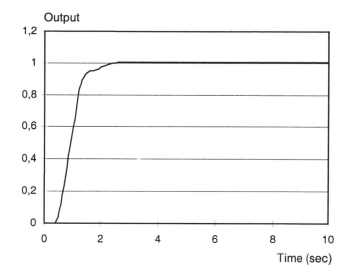

Fig. 2.16 (a) Response of plant using FLC (a=0.4; other process parameters as given by equation (2.25))

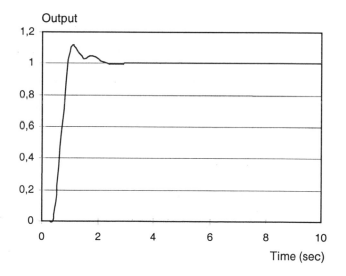

Fig. 2.16 (b) Response of plant using FLC (a=0.2)

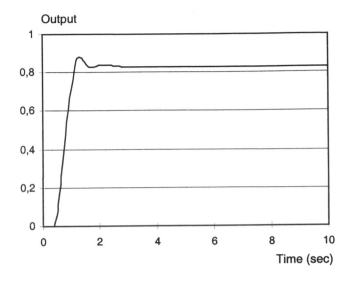

Fig. 2.16 (c) Response of plant using FLC (b=1.1)

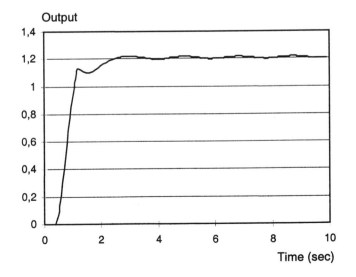

Fig. 2.16 (d) Response of plant using FLC (b=0.9)

Effect of Main Design Parameters on the Performance: Clearly, employing different performance indices for measuring the performance of controllers designed by the GA would lead to different FLC designs even if the same control parameters for the GA were used since the fitness value is derived from the performance indices. Figures 2.17(a-d) show the responses obtained for four different performance indices using the same GA parameters.

The dimensions of the relation matrix also affect the performance of the controller. A number of simulation experiments were conducted with different matrix sizes. The results are illustrated in Table 2.3. It was observed that too large a matrix made the search very slow. On the other hand, if the size of the relation matrix was too small, then it might not be possible to code the control strategy for the given process onto such a small matrix. Thus, there was an optimal size for the matrix. In the simulation studies, the size was maintained at 7×11 as mentioned earlier.

For the defuzzification procedure, two different methods were tried: the Average of Maxima and the Centre of Area methods. The responses obtained using these methods are given in Figures 2.18(a) and (b), respectively. In all subsequent studies the second method was adopted.

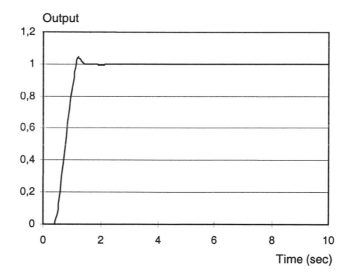

ITAE: Integral of Time Multiplied by Absolute Error

Fig. 2.17 (a) Response of plant using FLC designed according to ITAE criterion

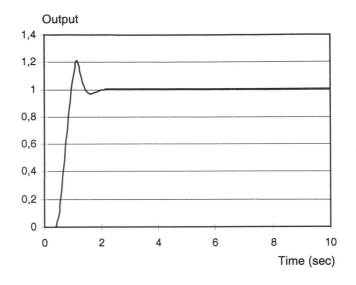

ITSE: Integral of Time Multiplied by Squared Error

Fig. 2.17 (b) Response of plant using FLC designed according to ITSE criterion

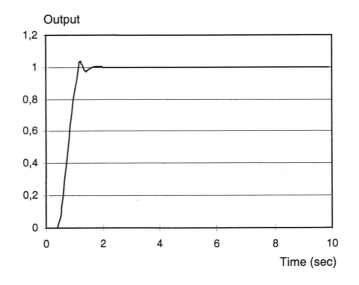

IAE: Integral of Absolute Error

Fig. 2.17 (c) Response of plant using FLC designed according to IAE criterion

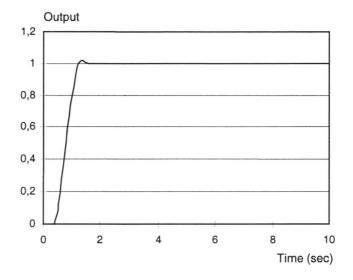

ISRAE: Integral of Square-Rooted Absolute Error

Fig. 2.17 (d) Response of plant using FLC designed according to ISRAE criterion

Table 2.3 Effect of size of relation matrix

Size of Matrix	ITAE
7×3	5.46
7×7	3.65
7×9	3.11
7×11	3.21
7×13	3.74
7×15	3.56
7×19	3.49
11×11	2.79

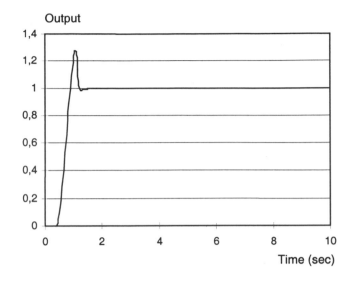

Fig. 2.18 (a) Response of plant using Average of Maxima criterion for defuzzification

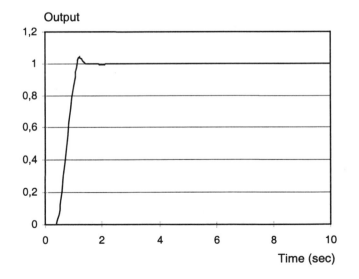

Fig. 2.18 (b) Response of plant using Centre of Area method for defuzzification

Design of a PI Type Fuzzy Logic Controller. As seen from Figures 2.16(c-d) the "proportional" FLC proposed in the previous section had poor performance in the case of process parameter variation. Although reasonably robust against noise, it was not able to draw the actual output of the plant to the desired level. A fuzzy Proportional-plus-Integral (PI) controller would overcome such steady state errors. A pure fuzzy PI controller design using relation matrices requires at least 2 matrices [Gupta *et al.*, 1986], one for the Proportional action and the other for the Integral action. Therefore, if the same numbers of quantisation levels for the input and output spaces are used, the computation time will be doubled when compared to the FLC with a proportional action only. A fuzzy PI controller is desired which can handle process parameter changes better than the original FLC but only requires a similar computation time. The structure of the proposed fuzzy PI controller is depicted in Figure 2.19. Since this controller is not a pure fuzzy PI controller, but a proportional FLC with integral action, it will be called a PI-type fuzzy controller.

In Figure 2.19, K_I is the coefficient of the integrator. This parameter can be either intuitively set by the control system designer or optimised using a GA, as with the relation matrix of the FLC. In this study, the latter method was adopted. The addition of the integral action to the FLC increased the number of parameters to be optimised by one. An individual in the string population thus represents the relation matrix of the FLC together with the parameter K_I for which eight bits were also devoted.

The results obtained using the proposed controller are shown in Figures 2.20, 2.21 and 2.22(a-d). Figure 2.20 presents the step response of the plant defined by equations (2.24) and (2.25). Figures 2.22(a-d) show the responses obtained from the plant after some of its parameters were changed. The response of the system in the presence of white noise with variance equal to 0.0132 is given in Figure 2.21. The value of K_I as obtained by the GA was 0.115.

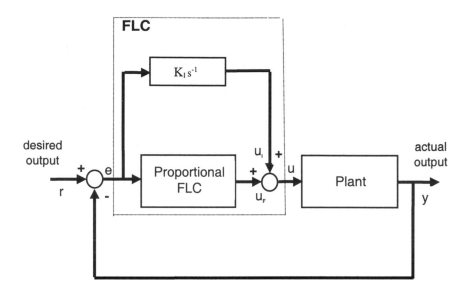

Fig. 2.19 Structure of proposed PI-type FLC

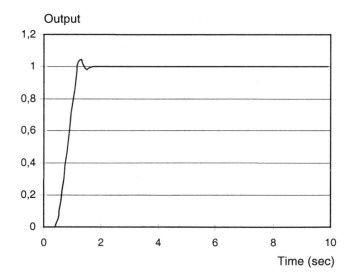

Fig. 2.20 Response of plant using PI-type FLC

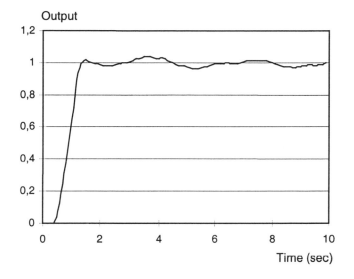

Fig. 2.21 Response of plant using fuzzy PI-type controller (white noise present)

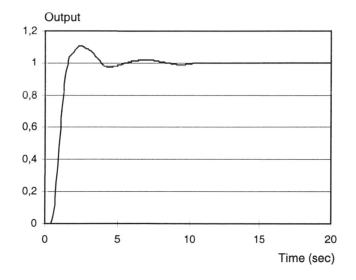

Fig. 2.22 (a) Response of plant using fuzzy PI-type controller (a=0.4)

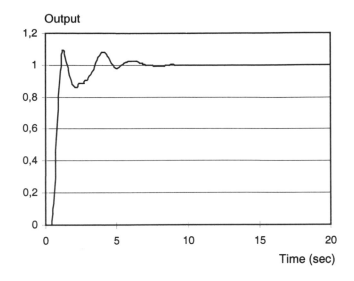

Fig. 2.22 (b) Response of plant using fuzzy PI-type controller (a=0.2)

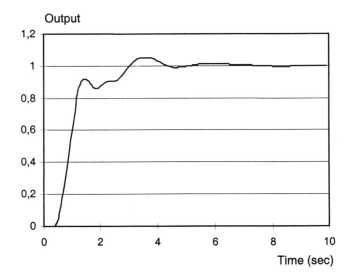

Fig. 2.22 (c) Response of plant using fuzzy PI-type controller (b=1.1)

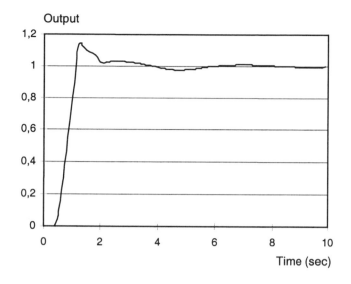

Fig. 2.22 (d) Response of plant using fuzzy PI-type controller (b=0.9)

2.2.2 Training Recurrent Neural Networks

As mentioned in Chapter 1, from a structural point of view, there are two main types of neural networks: feedforward and recurrent neural networks. The interconnections between the neurons in a feedforward neural network (FNN) are such that information can flow only in one direction, from input neurons to output neurons. In a recurrent neural network (RNN), two types of connections, feedforward and feedback connections, allow information to propagate in two directions, from input neurons to output neurons and vice versa.

FNNs have been successfully applied to the identification of dynamic systems [Pham and Liu, 1997]. However, they generally require a large number of input neurons and thus necessitate a long computation time as well as having a high probability of being affected by external noise. Also, it is difficult to train a FNN to act as an independent system simulator [Pham and Liu, 1997].

RNNs have attracted the attention of researchers in the field of dynamic system identification since they do not suffer from the above problems [Pham and Liu, 1996; Ku and Lee, 1995; Pham and Liu, 1992]. Two simple types of RNNs are the Elman network [Elman, 1990] and the Jordan network [Jordan, 1986]. Modified versions of these RNNs have been developed and their performance in system identification has been tested against the original nets by the authors' group [Pham and Liu, 1992; Pham and Oh, 1992]. In that work, the training of the RNNs was implemented using the simple backpropagation (BP) algorithm [Rumelhart, 1986]. To apply this algorithm, the RNNs were regarded as FNNs, with only the feedforward connections having to be trained. The feedback connections all had constant pre-defined weights, the values of which were fixed experimentally by the user.

This section presents the results obtained in evaluating the performance of the original Elman and Jordan nets and the modified nets when trained by a GA for dynamic system identification. First the structures of these nets are reviewed. Second the way in which the GA was used to train the given nets to identify a dynamic system is explained. Finally the results obtained for three different Single-Input Single-Output (SISO) systems are given.

Elman Network and Jordan Network. Figure 2.23 depicts the original Elman network which is a neural network with three layers of neurons. The first layer consists of two different groups of neurons. These are the group of external input neurons and the group of internal input neurons also called context units. The inputs to the context units are the outputs of the hidden neurons forming the second or hidden layer. The outputs of the context units and the external input neurons are fed to the hidden neurons. Context units are also known as memory units as they store the previous output of the hidden neurons. The third layer is simply composed of output neurons.

Although, theoretically, an Elman network with all feedback connections from the hidden layer to the context layer set to 1 can represent an arbitrary nth order system, where n is the number of context units, it cannot be trained to do so using the standard BP algorithm [Pham and Liu, 1992]. By introducing self-feedback connections to the context units of the original Elman network and thereby increasing its dynamic memory capacity, it is possible to apply the standard BP algorithm to teach the network that task [Pham and Liu, 1992]. The modified Elman network is shown in Figure 2.24. The values of the self-connection weights are fixed between 0 and 1 before starting training.

The idea of introducing self-feedback connections for the context units was borrowed from the Jordan network shown in Figure 2.25. This neural network also has three layers, with the main feedback connections taken from the output layer to the context layer. It has been shown theoretically [Pham and Oh, 1992] that the original Jordan network is not capable of representing arbitrary dynamic systems. However, by adding the feedback connections from the hidden layer to the context layer, similarly to the case of the Elman network, a modified Jordan network (see Figure 2.26) is obtained that can be trained using the standard BP algorithm to model different dynamic systems [Pham and Oh, 1992]. As with the modified Elman network, the values of the feedback connection weights have to be fixed by the user if the standard BP algorithm is employed.

Training neural network. Genetic algorithms have been used in the area of neural networks for three main tasks: training the weights of connections, designing the structure of a network and finding an optimal learning rule [Chalmers, 1990]. The first and second tasks have been studied by several researchers with promising results. However, most of the studies to date have been on feedforward neural networks rather than recurrent networks [Whitely *et al.*, 1989; Harp *et al.*, 1991].

In this section, the GA is used to train the weights of recurrent neural networks, assuming that the structure of these networks has been decided. That is, the number of layers, the type and number of neurons in each layer, the pattern of connections, the permissible ranges of trainable connection weights, and the values of constant connection weights, if any, are all known. Here, again the GA presented in Figure 1.12 is employed. The solutions in a population represent possible RNNs. RNNs are evaluated and improved by applying genetic operators.

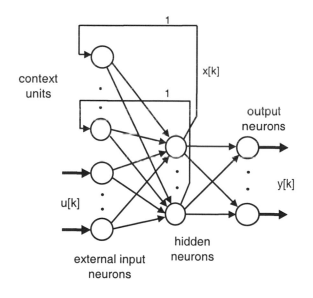

Fig. 2.23 Structure of the original Elman network

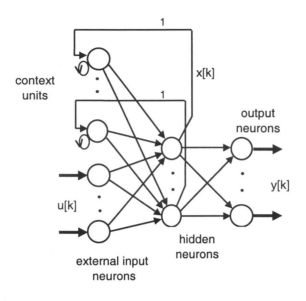

Fig. 2.24 Structure of the modified Elman network

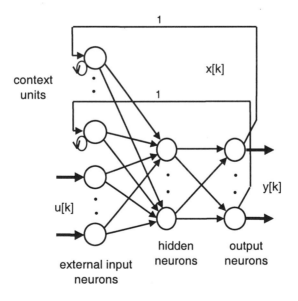

Fig. 2.25 Structure of the original Jordan network

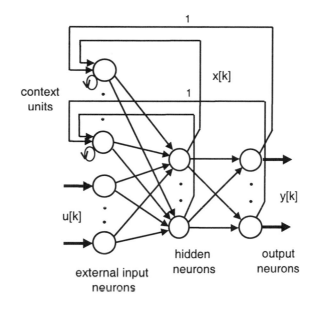

Fig. 2.26 Structure of the modified Jordan network

Each solution in the initial and subsequent populations is a string comprising n elements, where n is the number of trainable connections (see Figure 2.27). Each element is a 16-bit binary string holding the value of a trainable connection weight. It was found that sixteen bits gave adequate resolution for both the feedforward connections (which can have positive or negative weights) and the feedback connections (which have weight values in the range 0 to 1). A much shorter string could cause the search to fail due to inadequate resolution and a longer string could delay the convergence process unacceptably. Note that from the point of view of the GA, all connection weights are handled in the same way. Training of the feedback connections is carried out identically to training of the feedforward connections, unlike the case of the commonly used BP training algorithm.

The use of a GA to train a RNN to model a dynamic plant is illustrated in Figure 2.28. A sequence of input signals $u(k)$, ($k = 0, 1, ...$), is fed to both the plant and the RNN (represented by a solution string taken from the current population). The output signals $y_p(k+1)$ for the plant and $y_m(k+1)$ for the RNN are compared and the differences $e(k+1) = |y_p(k+1)-y_m(k+1)|$ are computed. The sum of all $e(k+1)$ for the whole sequence is used as a measure of the goodness or fitness of the particular RNN under consideration. These computations are carried out for all the networks of the current population. After the aforementioned genetic operators are applied, based on the fitness values obtained, a new population is created and the above procedure repeated. The new population normally will have a greater average fitness than preceding populations and eventually a RNN will emerge having connection weights adjusted such that it correctly models the input-output behaviour of the given plant.

W_{ij} :weight of connection from neuron j to neuron i

Fig. 2.27 Representation of the trainable weights of a RNN in string form

The choice of the training input sequence is important to successful training. Sinusoidal and random input sequences are usually employed [Pham and Oh, 1992]. In this investigation, a random sequence of 200 input signals was adopted. This number of input signals was found to be a good compromise between a lower number which might be insufficient to represent the input-output behaviour of a plant and a larger number which lengthens the training process.

Simulation Results. Simulations were conducted to study the ability of the above four types of RNNs to be trained by a GA to model three different dynamic plants. The first plant was a third-order linear plant. The second and third plants were non-linear plants. A sampling period of 0.01 seconds was used in all cases.

Plant 1. This plant had the following discrete input-output equation:

$$y(k+1) = 2.627771\ y(k)-2.333261\ y(k-1) + 0.697676\ y(k-2)$$
$$+ 0.017203\ u(k) - 0.030862\ u(k-1) + 0.014086\ u(k-2) \qquad (2.28)$$

The modelling was carried out using both the modified Elman and Jordan networks with all linear neurons. The training input signal, $u(k)$, $k = 0,1,...,199$, was random and varied between -1.0 and +1.0. First, results were obtained by assuming that only the feedforward connections were trainable. The responses from the plant and the modified Elman and Jordan networks are presented in Figures 2.29(a) and (b). The networks were subsequently trained with all connections modifiable. The responses produced are shown in Figures 2.30(a) and (b).

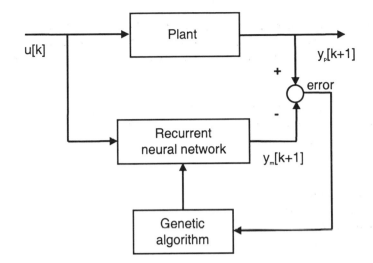

Fig. 2.28 Scheme for training a recurrent neural network to identify a plant

Plant 2. This was a non-linear system with the following discrete-time equation:

$$y(k+1) = y(k) / (1.5 + y^2(k)) - 0.3\ y(k-1) + 0.5\ u(k) \tag{2.29}$$

The original Elman network and the modified Elman and Jordan networks with non-linear neurons in the hidden layer and linear neurons in the remaining layers were employed. The hyperbolic tangent function was adopted as the activation function of the non-linear neurons. The neural networks were trained using the same sequence of random input signals as mentioned previously. The responses obtained using the networks, taking only the feedforward connections as variable, are presented in Figures 2.31(a), (b) and (c), respectively. The results produced when all the connections were trainable are given in Figures 2.32(a), (b) and (c).

Plant 3. This was a non-linear system with the following discrete-time equation:

$$\begin{aligned} y(k+1) = (1.752821\ y(k) - 0.818731\ y(k-1) + 0.011698\ u(k) \\ + 0.010942\ u(k-1)) / (1+y^2(k-1)) \end{aligned} \tag{2.30}$$

The original Elman network with non-linear neurons in the hidden layer and linear neurons in the remaining layers, as in the case of Plant 2, was employed. The hyperbolic tangent function was adopted as the activation function of the non-linear neurons. The responses obtained using the network, taking only the feedforward connections as variable, are presented in Figure 2.33. The result produced when all the connections were trainable is given in Figure 2.34. It can be seen that a very accurate model of the plant could be obtained with this Elman network and therefore the more complex modified Elman and Jordan networks were not tried.

The Mean-Square-Error (MSE) values computed for networks with constant feedback connection weights and all variable connection weights are presented in Tables 2.4(a) and (b) respectively.

In all cases, when only the feedforward connection weights were modifiable, the GA was run for 10000 generations and, when all the connection weights could be trained, the algorithm was implemented for 3000 generations. The other GA control parameters [Grefenstette, 1986] were maintained in all simulations at the following values:

Population size : 50
Crossover rate : 0.85
Mutation rate : 0.01
Generation gap : 0.90

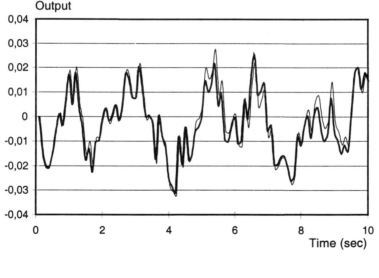

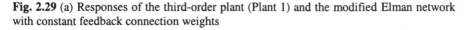

Fig. 2.29 (a) Responses of the third-order plant (Plant 1) and the modified Elman network with constant feedback connection weights

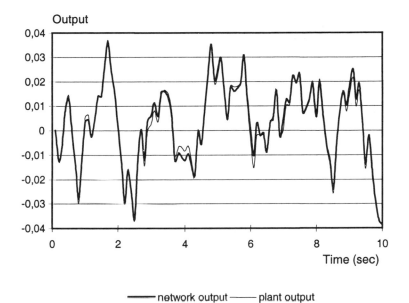

Fig. 2.29 (b) Responses of the third-order plant (Plant 1) and the modified Jordan network with constant feedback connection weights

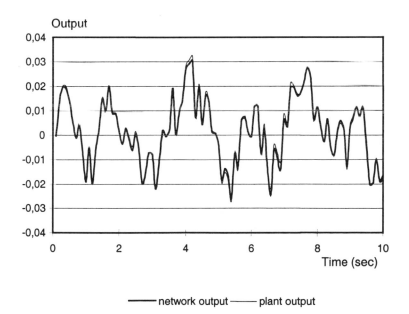

Fig. 2.30 (a) Responses of the third-order plant (Plant 1) and the modified Elman network with all variable connection weights

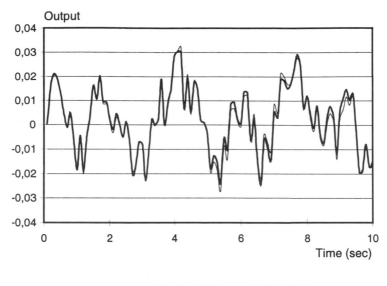

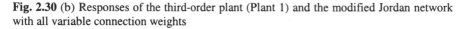

Fig. 2.30 (b) Responses of the third-order plant (Plant 1) and the modified Jordan network with all variable connection weights

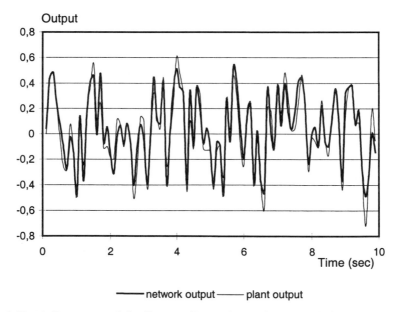

Fig. 2.31 (a) Responses of the first non-linear plant (Plant 2) and the original Elman network with constant feedback connection weights

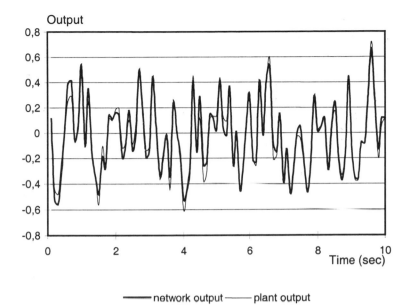

Fig. 2.31 (b) Responses of the first non-linear plant (Plant 2) and the modified Elman network with constant feedback connection weights

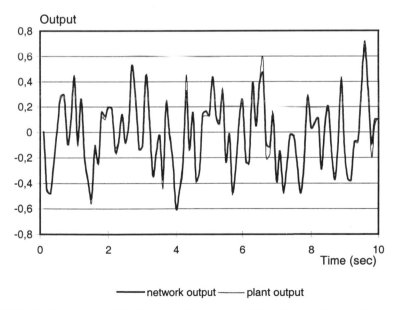

Fig. 2.31 (c) Responses of the first non-linear plant (Plant 2) and the modified Jordan network with constant feedback connection weights

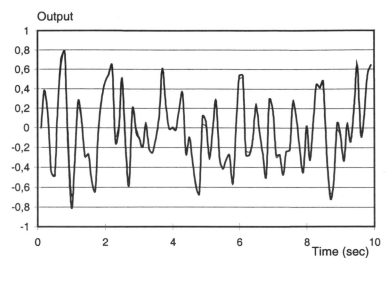

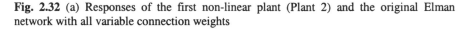

Fig. 2.32 (a) Responses of the first non-linear plant (Plant 2) and the original Elman network with all variable connection weights

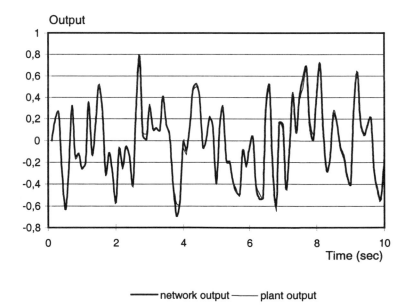

Fig. 2.32 (b) Responses of the first non-linear plant (Plant 2) and the modified Elman network with all variable connection weights

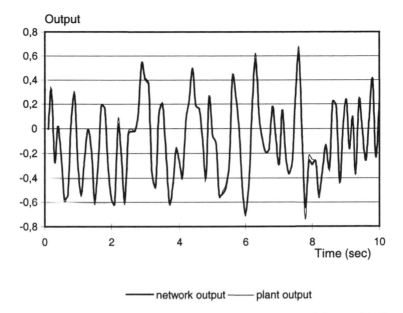

network output —— plant output

Fig. 2.32 (c) Responses of the first non-linear plant (Plant 2) and the modified Jordan network with all variable connection weights

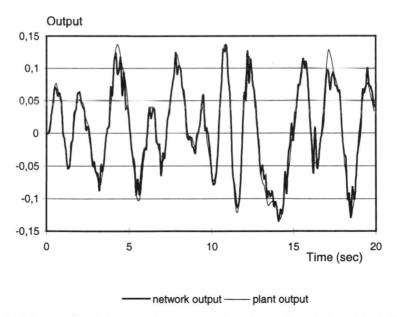

network output —— plant output

Fig. 2.33 Responses of the second non-linear plant (Plant 3) and the original Elman network with constant feedback connection weights

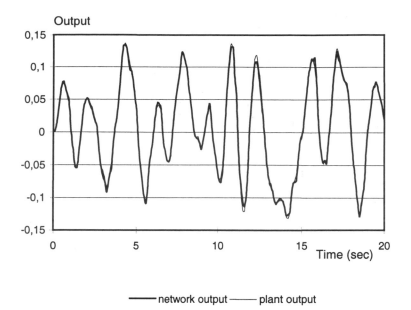

Fig. 2.34 Responses of the second non-linear plant (Plant 3) and the original Elman network with all variable connection weights

Additional Tests and Discussion. Experiments were also conducted for both first and second-order linear plants although the results are not presented here. The first-order plant was easily identified using the original Elman network structure. However, as expected from the theoretical analysis given in [Pham and Oh, 1992], the original Jordan network was unable to model the plant and produced poor results. Both the original and modified Elman networks could identify the second-order plant successfully. Note that an original Elman network with an identical structure to that adopted for the original Elman network employed in this work and trained using the standard BP algorithm had failed to identify the same plant [Pham and Liu, 1992].

The third-order plant could not be identified using the original Elman network but was successfully modelled by both the modified Elman and Jordan networks. This further confirms the advantages of the modifications [Pham and Liu, 1992; Pham and Oh, 1992].

With Jordan networks, training was more difficult than with Elman networks. For all network structures, the training was significantly faster when all connection weights were modifiable than when only the feedforward connection weights could be changed. This was probably due to the fact that in the former case the GA had more freedom to evolve good solutions. Thus, by using a GA, not only was it possible and simple to train the feedback connection weights, but the training time required was lower than for the feedforward connection weights alone.

A drawback of the GA compared to the BP algorithm is that the GA is inherently slower as it has to operate on the weights of a population of neural networks whereas the BP algorithm only deals with the weights of one network.

Table 2.4 (a) MSE values for networks with constant feedback connection weights
(b) MSE values for networks with all variable connection weights

Plant	Original Elman	Modified Elman	Modified Jordan
1		0.0001325	0.0002487
2	0.0018662	0.0012549	0.0006153
3	0.0002261		

(a)

Plant	Original Elman	Modified Elman	Modified Jordan
1		0.0000548	0.0000492
2	0.0005874	0.0004936	0.0003812
3	0.0000193		

(b)

2.2.3 Adaptive Fuzzy Logic Controller Design

A fuzzy logic controller (FLC) which cannot adapt to changes in process dynamics and disturbances is a static controller. In the control of a process where large parameter changes occur or little process knowledge is available, an adaptive controller is required to achieve good performance.

The first FLC with an adaptation ability was developed by Procyk and Mamdani [Procyk and Mamdani, 1979]. This controller has been studied and improved by other researchers [Yamazaki, 1982; Sugiyama, 1986]. It has been successfully applied in cases where a small amount of a priori knowledge is available to design the controller [Shao, 1988; Linkens *et al.*, 1991]. The structure of this self-organising controller (SOC) is depicted in Figure 2.35.

The main idea underlying this controller is to modify its rule base according to its closed-loop performance. An adaptive controller must be able to measure its own performance and change its control strategy as a result. In a SOC, a performance table is employed to determine the amount of modification to the controller depending on the measured performance. Performance is defined by the difference between the actual and desired outputs of the plant under control. If it is poor, correction is made to the rules responsible using the performance table and the measured performance data.

As described in [Narendra *et al.*, 1989], for example, there are mainly two types of adaptive controllers: model reference adaptive controllers (MRAC) and self-tuning regulators (STR). The structures of these controllers are given in Figures 2.36 and 2.37, respectively. In terms of the structure and operation, a SOC is similar to an MRAC. Both can be regarded as direct adaptive systems which do not require obtaining an updated model of the process to modify the controller. In an MRAC, a reference model is required to calculate the performance of the controller, a task which is carried out in a SOC using a performance table. The main problem with a SOC is its dependence on the performance table. It is difficult to develop an optimal performance table and there is no systematic way to produce one for a specific problem.

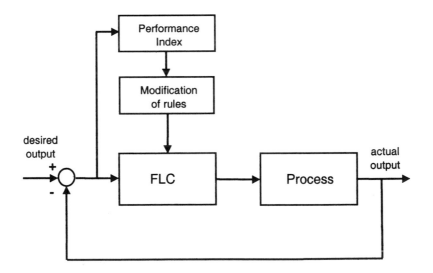

Fig. 2.35 Basic structure of a simple self-organising FLC

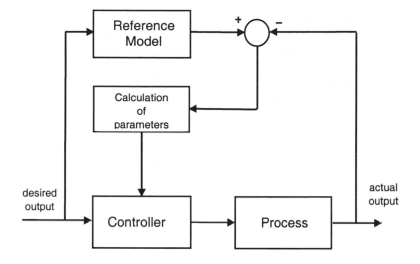

Fig. 2.36 Basic structure of a Model Reference Adaptive Control system

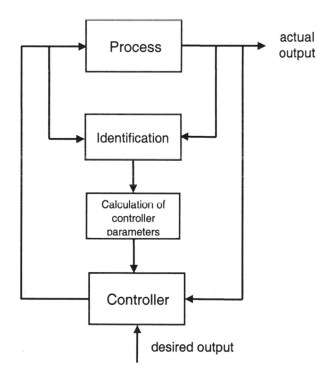

Fig. 2.37 Basic structure of a self-tuning regulator

The second type of adaptive controller, the self-tuning regulator, does not use a reference model or performance table to evaluate performance. It consists of two main parts: on-line identification of the process and modification of the controller depending on the latest process model identified. In other words, it employs an intermediate model of the process to tune the controller. Consequently, this type controller is also called an indirect adaptive controller.

An adaptive FLC based on the principles of self-tuning regulators would not suffer from the problems of a SOC. Adaptive fuzzy logic control schemes based on the self-tuning regulator principle have been also proposed [Moore *et al.*, 1992]. In the work described in [Moore *et al.*, 1992], an adaptive fuzzy model is constructed of the process and a fuzzy logic controller is designed by inverting the fuzzy model. This method hinges upon obtaining an accurate process model which is not simple to achieve.

This section describes a new adaptive fuzzy logic control (AFLC) scheme. The proposed scheme is based on the structure of the self-tuning regulator and employs neural network and GA techniques. First the scheme is described. Second simulation results obtained for the control of linear and non-linear processes are presented. Finally the results are discussed.

Proposed Adaptive FLC. The structure of the proposed self-tuning adaptive fuzzy logic control system is given in Figure 2.38. As with the self-tuning regulator of Figure 2.37, the proposed system comprises two main parts: on-line identification of the process and modification of the FLC using the identified model.

Identification. To achieve a high-performance AFLC, the process has to be accurately identified. The identification scheme adopted in this work was based on a neural network that had been shown capable of producing accurate results for different types of processes in Section 2.2.2.

Neural Network Structure: Figure 2.39 presents the structure of the neural network used. The network is a modified version of the recurrent network proposed by Elman [1990]. The self-feedback connections of the context neurons, which are added to the original version of the Elman network, increase the memorising ability of the network and enable it to be trained more easily. The outputs of the context neurons and external input neurons are fed to the hidden neurons.

The numbers of external input and output neurons of the network are dictated by the numbers of process inputs and outputs. The numbers of context and hidden neurons are related to the order of the process.

In this work, as the process to be identified and controlled was a single-input-single-output process, a network with one external input and one output was adopted. The numbers of context and hidden neurons were chosen as 6 each, after experimentation with different network sizes. This gave a total of 60 connection weights.

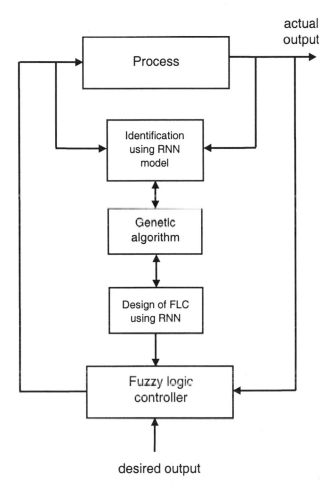

Fig. 2.38 Structure of the proposed adaptive FLC system

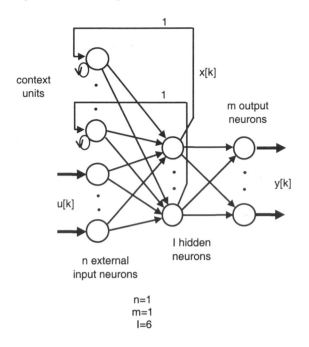

Fig. 2.39 Structure of the RNN used for plant identification

Neural Network Training: A GA was adopted for training the network. The GA used is the one described in Figure 1.12. Sixteen bits were dedicated to coding the value of each weight of the neural network. This bit length was adopted as a compromise between a higher resolution, which would slow the identification operation, and a lower resolution, which would produce inaccurate results. Using 16 bits per weight yields a total length of 960 bits for a chromosome representing all the network weights.

Before starting the adaptive control of the process, an initial neural network model was required for designing the FLC used at the beginning of the control operation. This model was obtained by training the network off-line with a set of input-output data pairs collected from the process. An approximate mathematical model of the process could equally have been employed to generate the initial training data.

During on-line control, the neural network model was modified after each sampling period by training the network with a set of data obtained from the process to minimise a performance index defined as:

$$J_m = \sum_{k=1}^{w} (y_p[k] - y_m[k])^2 \tag{2.31}$$

In equation (2.31), w is the number of input-output data pairs in the buffer containing the data set, y_p is the plant output and y_m is the network output. $y_m[k]$ and $y_p[k]$ correspond to the same input data applied to the process at sampling time k.

The data buffer operated as a first-in-first-out stack. That is, the oldest data pair was removed each time a new data pair was added to the buffer. This updating process normally took place at each sampling time. However, when the desired output was constant and there was no noise and no change in the dynamics of the process, the data pairs stored in the buffer became identical to one another after a certain time. Such a data set could not be used to train the neural network to model the process. To avoid this problem, the most recently sampled data pair was compared to the data pair last added to the buffer and, if they were the same, the latest sample was discarded. In other words, the data set used for training was not changed and training of the network was suspended as the process was considered to have been modelled with a sufficient degree of accuracy.

The number of data pairs in the buffer was related to the order of the process and the time delay. If too many data pairs were used, the adaptation of the model to changes in the process would be slow. This is because, according to the performance criterion adopted, at least half of the buffer must contain data samples obtained following a parameter change before the effect of the change was felt. If too few data pairs were employed, then it might not be possible to identify the process correctly. In this work, 200 data pairs were used. This figure was again adopted as a compromise between adaptation speed and identification accuracy.

The fitness of each network generated by the GA was obtained from the performance index J_m as :

$$fit_i = (J_m + C_m)^{-1} \tag{2.32}$$

where C_m is a small constant (0.001) added to J_m to avoid (the unlikely case of) dividing by zero.

After a given number of generations, the GA was stopped and the network with the highest fitness value was taken as the best model of the process and used in the design of the FLC. The maximum number of generations at which the GA was stopped was determined according to the sampling period of the control system. For a large sampling period, this number could be set high. Otherwise, a small number had to be selected. Using a large number of generations increased the possibility of finding a good model. In this work, this number was taken as 10 which means that after a new data pair was added to the buffer, the GA was executed for 10 generations to update the current model.

Adaptation Speed Increase: As mentioned in the previous section, although the neural network model was re-trained after each sampling period, the effect of changes was not felt until after at least half of the data buffer was filled with new samples. This problem, which was due to the performance index adopted caused the adaptation speed to be slow.

As seen from equation (2.31), in the evaluation of the performance of a network model, the weighting factors of all errors were equal to 1. Assigning a different weighting factor to each error, depending on the sampling time k, could increase the adaptation speed of the model. In this case, equation (2.31) can be written as:

$$J_m = \sum_{k=1}^{w} \xi(k) \; (y_p[k]-y_m[k])^2 \tag{2.33}$$

where $\xi(k)$ is the weighting factor of the error at time k.

Different functions can be employed for the weighting factor values $\xi(k)$. In this work,

$$\xi(k) = \sqrt{(k/w)} \tag{2.34}$$

Using this function, the range of weighting factors is 0.071 (k=1, w=200) to 1 (k= 200, w=200). This means that the most recent error is assigned a weighting factor of 1 and the oldest error, a weighting factor of 0.071 to calculate the performance of a neural network model.

Controller Design. After the best model for the process had been identified, the evolutionary controller design procedure was carried out (see Section 2.2.1). The FLC that was best able to control the model was then employed to control the real process.

As with the identification operation, the GA used for obtaining the relation matrix of the controller was executed for 10 generations to find the best relation matrix.

To evaluate the performance of a given FLC, a step input was applied to the system consisting of the FLC and the model. The performance was computed as:

$$J_c = \sum_{i=1}^{L} i\, T \, |y_d[i] - y_m[i]| \tag{2.35}$$

where, J_c is the performance index of the controller, y_d is the desired output of the process, y_m is the actual output of the model under control, L is the simulation length and T is the sampling period.

The fitness of an FLC was calculated as:

$$fit_i = (J_c + C_c)^{-1} \tag{2.36}$$

where C_c is a small constant (0.01) added to J_c again as a precautionary measure to avoid dividing by zero.

Results. The above scheme was tested in simulation on two different sampled-data processes. The first process was a linear second order process and the second process was non-linear. The sampling period in both cases was 0.01 seconds.

Plant 1: The discrete input-output equation for this process was:

$$y(k+1) = A_1\, y(k) + A_2\, y(k\text{-}1) + B_1\, u(k) + B_2\, u(k\text{-}1) \tag{2.37}$$

where $A_1 = 0.72357$, $A_2 = -0.70482$, $B_1 = 0.0090$ and $B_2 = 0.0082$.

White noise ε was added to the output of the process, so that the discrete input-output equation became:

$$y(k+1) = A_1\, y(k) + A_2\, y(k\text{-}1) + B_1\, u(k) + B_2\, u(k\text{-}1) + \varepsilon(k+1) \tag{2.38}$$

The training input signal during the initial training carried out before starting the control of the process was pseudo random, varying between −1.0 and +1.0. After this initial training phase, which lasted for 300 generations, the data pairs were obtained from the output of the FLC and the output of the plant.

The response of the noise-free process defined by equation (2.37) under the control of an adaptive FLC in the case of process parameter variations is given in

Figure 2.40. At $t = 1.5$ ($k = 150$), the process parameters were changed to $A_1 = 1.693778$, $A_2 = -0.786628$, $B_1 = 0.013822$ and $B_2 = 0.012756$.

Figure 2.41 shows the response of the system when output noise with variance equal to 0.0058 was present. Figure 2.42 depicts the response of the system for the combined case of output noise (variance = 0.0058) and parameter variations. Here, the parameters were again changed at $t = 1.5$ seconds.

Plant 2: This non-linear process had the following discrete input-output equation:

$$y(k+1) = A\, y(k) / [B + C\, y^2(k)] - D\, y(k\text{-}1) + E\, u(k) + \varepsilon(k+1) \qquad (2.39)$$

Here, $A = 1$, $B = 1.5$, $C = 1$, $D = 0.3$ and $E = 0.5$. As before, the neural model was initially trained for 300 generations, using pseudo random inputs between -1.0 and +1.0 before starting the control process.

To test the system when the process parameters vary, at $t = 1.5$ seconds, the parameters of the process were changed from the above values to: $A = 1.1$, $B = 1.75$, $C = 1.5$, $D = 0.4$ and $E = 0.4$. The response of the system is shown in Figure 2.43. The response of the original system when white noise with variance equal to 0.0137 was present at the output is given in Figure 2.44.

Figure 2.45 shows the response of system for the combined case of output noise (variance = 0.0137) and parameter variations. (As before, the parameters were changed at $t = 1.5$ seconds).

Figures 2.46 and 2.47 show the responses of the adaptive system to the parameter changes for the linear and non-linear processes, respectively, when the strategy suggested in Section 2.1.3 was used for calculating the performance of the neural network model of the process.

Discussion. The results obtained have clearly shown that the on-line adaptation was effective. Note the reduction in the overshoot and steady state errors in Figures 2.40, 2.42, 2.43, 2.46 and 2.47 after the system had adapted to the parameter changes.

The processes employed in this study were the same as those described by Oh [1993] who investigated their control using an adaptive scheme based completely on neural networks.

As shown in Table 2.5, the performance of the proposed adaptive fuzzy logic control scheme is superior to that of the scheme proposed in [Oh, 1993] even without using the adaptation speed increase strategy.

The strategy suggested in Section 2.1.3 for the computation of the performance of a network model significantly increased the adaptation speed of the system. As seen from Figures 2.40, 2.43, 2.46 and 2.47, the improvement in the adaptation speed was more than 30%.

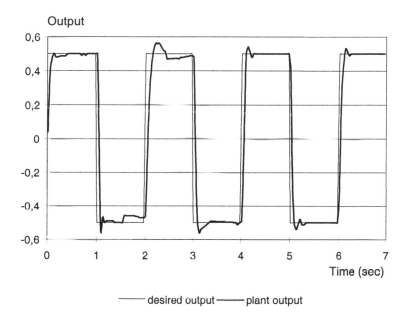

Fig. 2.40 Response of the control system with parameter changes (Plant 1)

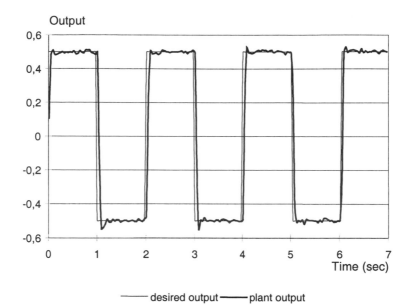

Fig. 2.41 Response of the control system with noise (Plant 1)

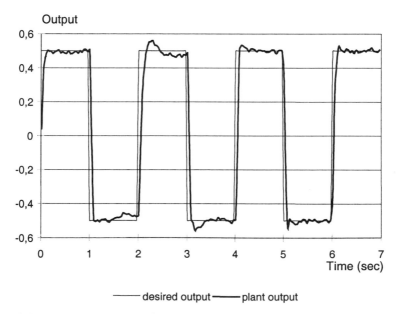

Fig. 2.42 Response of the control system with noise and parameter changes (Plant 1)

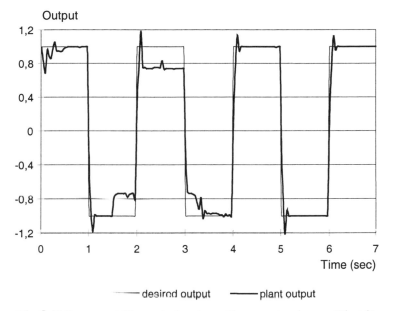

Fig. 2.43 Response of the control system with parameter changes (Plant 2)

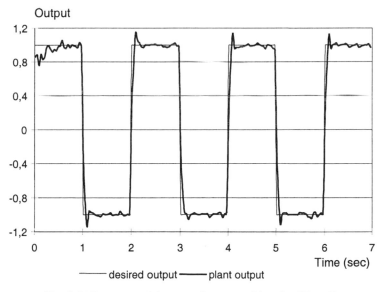

Fig. 2.44 Response of the control system with noise (Plant 2)

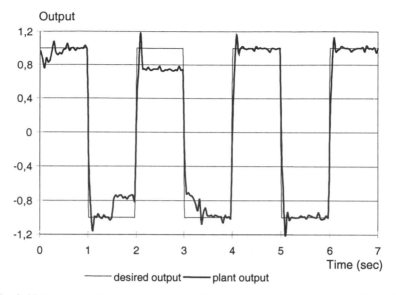

Fig. 2.45 Response of the control system with parameter changes and noise (Plant 2)

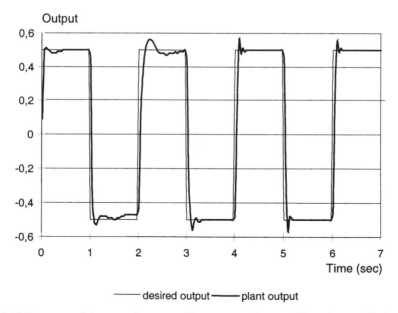

Fig. 2.46 Response of the control system with parameter changes (Plant 1 – modified performance measure)

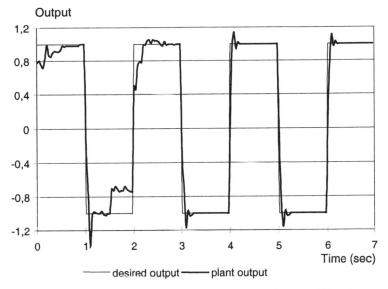

Fig. 2.47 Response of the control system with parameter changes (Plant 2 – modified performance measure)

Table 2.5 Performances of the proposed scheme and Oh's scheme

scheme	plant	variance of noise	normalised MSE	adaptation time (sec)
proposed	linear	0.0	0.06157	2.5
		0.0058	0.06824	2.5
	non-linear	0.0	0.06384	2
		0.0137	0.08016	2
Oh's	linear	0.0	0.09476	192
		0.0028	0.07724	392
	non-linear	0.0	0.14129	142
		0.0142	0.14433	142

2.2.4 Preliminary Gearbox Design

Problem Statement. Figure 2.48 shows an example of a gearbox as a device for transmitting power between a motor and a conveyor.

Assumptions: For simplicity, several assumptions have been made:

(a) Only one input speed is chosen for the gearbox.
(b) Only one output speed is required from the gearbox.
(c) Only parallel shafts are used in the gearbox.
(d) The power transmission between two adjacent shafts is accomplished by only one pair of gears.

The task of the GA is to find as many alternative gearbox designs as possible that are capable of approximately producing the required output speed using one of the available input speeds (it is assumed that designers have a small number of alternative motors at their disposal). In this application, the four techniques suggested in section 2.1.3 have been employed to help a GA to find multiple solutions. The main advantage of these techniques is that they all encourage genetic variety, and hence the production of different solutions, without disrupting the gene population.

Parameters: The following are therefore known parameters:

(a) The speed of the motor (700, 1000, 1500 or 3000 r/min);
(b) The required output speed (specified in revolutions per minute by the program user).

The following parameters are to be computed by the GA:

(a) Number of shafts;
(b) Number of teeth for each gear.

Constraints:

(a) The minimum number of teeth for a gear is 18.
(b) The maximum number of shafts, including the input and output shafts, is 8 and the maximum number of gear pairs is 7.

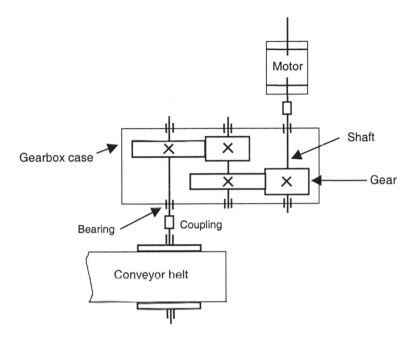

Fig. 2.48 An example of a gearbox

Genetic encoding. Binary strings encoding three types of parameters are used to represent a gearbox design: input speed option, number of shafts and number of teeth for each gear. Figure 2.49 shows a typical string. The total number of bits for the string is: 2+3+14×8 = 117

Evaluation Function. The factors considered in the construction of the evaluation fitness function, *F*, are listed below:

1. The number of shafts (N_s) generated must not exceed 8:

if $N_s > 8$ then $F = 0$

2. For a reduction gearbox, the number of teeth of the driven gear must be larger than that of the driving gear for each gear pair. Otherwise, the string is again considered invalid and its fitness *F* is given by:

$F = 28 - (2 \times WrongPair)$

where *WrongPair* is the number of invalid gear pairs.

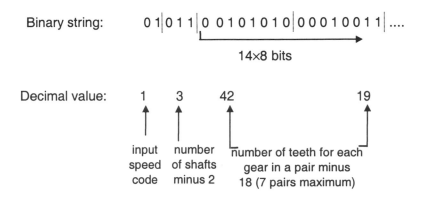

Fig. 2.49 Binary representation for gearbox design example

3. The transmission ratio achieved with one pair of gears must not exceed the maximum single-stage transmission ratio (MAXRATIO) by more than a given percentage. The allowed excess is 15 per cent. Excesses beyond this figure are accumulated and the total (*SumOfExcesses*) used to reduce the original string fitness is as follows:

$$F = F_{original} \ (1 - 0.1 SumOfExcesses)$$

4. In order to make the design compact, the number of shafts (*NoOfShafts*) should be as low as possible.

5. The difference between the total transmission ratio of the gearbox and the required ratio (*RatioDifference*) should be as small as possible.

6. The fitness value should be positive. However, for those valid strings satisfying constraints 1 to 3 above but having negative fitnesses, a fitness value is assigned that is slightly higher than the maximum attainable by an invalid string. The value used is 27. This is because those strings are still considered fitter than invalid ones.

Thus, the overall fitness function is as follows:

$$F = \begin{cases} 0, & \text{if } N_S > 8 \\ 28 - 2 \times WrongPair, & \text{if } WrongPair \neq 0 \\ \begin{pmatrix} MaxAllowedFitness \\ - \left[W1 \times 3^{(A1 \times NoOfShafts)} \\ + W2 \times 3^{(A2 \times RatioDifference)} \right] \end{pmatrix} \\ \times (1 - 0.1 \times SumOfExcesses), & \text{if } F \geq 0 \\ 27 & \text{if } F < 0 \end{cases} \tag{2.40}$$

where *MaxAllowedFitness*, set to 5000 in this problem, is a value based on the maximum obtainable fitness value of a design. *W1* and *W2* are penalty weights and *A1* and *A2* are coefficients. Experimentally *W1*, *W2*, *A1* and *A2* were tuned to yield a fitness function that was effective over a wide range of output speeds.

Note that there is no need to incorporate into the above fitness function constraints regarding the minimum number of teeth in a gear and the minimum number of shafts in the gearbox. This is because the design of the binary string representing a gearbox automatically excludes configurations not satisfying those constraints. (For example, the smallest value of the string section used to represent the number of shafts is '000', which yields (0+2) as the minimum shaft number.)

Distance Metric. The overall fitness value *F* was taken as the "reward" or "resource" to be shared. The transmission ratio of each gear pair of a solution was used to construct the distance metric d_{ij} :

$$d_{ij} = \sum_{k=1}^{P} \left| r_{ki} - r_{kj} \right| \tag{2.41}$$

where r_{ki} and r_{kj} are the transmission ratios for the *i*th and *j*th individuals in the string population and *P* is the number of gear pairs in them. When *P* is different for two individuals, they are regarded as being too far from each other for resource sharing.

Solution Acceptance Criteria and GA Performance Score. A "filtering" method was employed to capture all nonidentical satisfactory individuals generated during the evolution process. The total number of captured individuals was used to assess the performance of a particular implementation of the GA.

The "filtering" criteria for deciding whether a solution is acceptable are:

1. The output speed is within ±10 per cent of the required output speed.
2. The fitness value is higher than a lower limit (taken as approximately 0.2 per cent below the maximum fitness value).

Due to the very different performances of the GA at various output speeds, an indicator, namely the "combined score" (CS), has been employed to simplify the assessment and to give a uniform description of the merit of various modifications. CS is defined as:

$$CS = 100\left(\frac{\sum_{i=1}^{5} NS_i}{\sum_{i=1}^{5} NS_i^0} - 1 \right) \% \tag{2.42}$$

where NS_i^0 is the number of solutions from the original GA at output speed i and NS_i is the number of solutions from the modified GA at the same output speed. CS provides a comparison between the total numbers of solutions produced by the original GA and modified GAs at all of the five speeds tested (15, 30, 50, 75 and 100 r/min).

It is possible that a modified GA produces fewer solutions than the original GA at a certain speed. The number of output speeds at which the modified GA is inferior to the original GA is called NI.

Results and Discussion. Table 2.6 lists the CSs and NIs obtained from the GA when different modifications were made to it. The total number of solutions produced by the original GA = 1639. Clearly, as expected, the division-based "sharing" mechanism, either on its own or together with other modifications, did not yield good results. Note the negative values of CS and the high values of NI shown in Table 2.6.

The improvement to the GA when heuristics were applied from generation 0 is not significant (5 per cent). Delaying the application of heuristics by five generations gave a further increase in performance (14 per cent). When all the solution dispersal means including sharing, deflation, identical string elimination and delayed application of heuristics were combined, the modified GA produced 82 per cent more solutions than the original GA and 26 per cent more than the GA employing only sharing, deflation and identical string elimination.

Table 2.6 Results from GAs incorporating different modifications

	CS	NI
	%	
Original GA	0	0
GA+DBS	-27	3
GA+RBS	2	2
GA+SD	6	2
GA+ISE	0	0
GA+RBS+ISE+SD	44	0
GA+H from generation 0	5	2
GA+ H from generation 5	14	1
GA+RBS+ISE+SD+ H from generation 5	82	0
GA+DBS+ISE+SD+ H from generation 5	-97	5

DBS = division-based sharing
RBS = reduction-based sharing
ISE = identical string elimination
SD= solution deflation
H = heuristics

2.2.5 Ergonomic Workplace Layout Design

One area of design to which the discipline of ergonomics is being increasingly applied is that of ergonomic layout [Das and Grady, 1983]. Here the designer must determine the best locations for items or equipment within a working area. Typical problems might be to locate controls on a panel and items of test equipment in a factory workstation, or to find the optimum layout of furniture in an office. In each case a poor layout will result in low efficiency since the manipulation of items will be unnecessarily difficult [McCormick, 1982]. As a consequence, the operator may experience premature fatigue and, where safety is an issue, make dangerous mistakes.

A number of factors influence the designer in determining the most suitable locations for items in a workplace. In many cases, the most important factor will be the frequency with which each item is used. As a simple example, consider the design of a typical clerical office workstation. Here the operator requires the use of a word processor, an area for writing, a telephone, a printer and a fax machine. Depending on the type of clerical task, say, the word processor or the writing area might be the most frequently used facility. Thus, it is logical for the designer of the workstation to place one of them as close as possible to the operator's chair. The next most commonly used item may be the telephone. The designer should also place this close to the operator, but not at the expense of having to move the word processor or writing area. Finally, the printer and fax machine will typically be in less frequent use and so the designer can safely locate these items further away from the operator.

A second important factor influencing the designer is the interaction between objects. For example, it is generally convenient to place the printer near to the word processor to avoid excessive lengths of cable. Also, the telephone should be placed close to the word processor and writing area to allow the operator to access written information whilst talking.

Optimising the above factors is difficult for layouts with a large number of items. In many cases this will be beyond the ability of a human designer, and some kind of computational aid will be required. GAs have been applied to a wide range of areas, including symbolic layout [Fourman, 1985], scheduling [Davis, 1985] and search [Goldberg, 1988]. Preliminary work has applied them to ergonomic design [Pham and Onder, 1992].

This section describes the use of a GA to automate the process of optimising workplace layout. After defining a mathematical representation for the layout problem, the design factors described above are encoded into an evaluation function. A genetic algorithm is then used to manipulate the location of items in the layout to minimise this evaluation function. Results are given for a number of workplace layouts.

Mathematical Representation of the Workplace Layout Problem. The factors described in the previous section can be expressed as two simple design rules, namely to place the most frequently used items closest to the operator, and to place items which require a strong interaction close together. A simple form of the workplace layout problem can be described as follows.

Consider a set of n equal size objects, which are placed at locations x_i, y_i (see Fig. 2.50). Assume that the operator is located at co-ordinates x_0, y_0. Determine the co-ordinates x_i, y_i for the n objects which best satisfy the above design rules.

In order to determine how well a particular layout satisfies the design rules, they can be encapsulated into an evaluation function which will yield an optimal layout when minimised. One possible form of the evaluation function is:

$$e = \frac{1}{2} \sum_{i=0}^{n} \sum_{j=0}^{n} w_{ij} d_{ij}^2 \qquad (2.43)$$

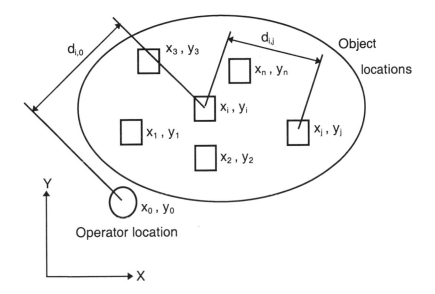

Fig. 2.50 Locations of objects in the workplace layout problem.

Here d_{ij} represents the distance between objects i and j, which can be determined using:

$$d_{ij} = \sqrt{(x_i - x_j)^2 + (y_i - y_j)^2} \qquad (2.44)$$

w_{ij} are weighting coefficients specified by the user, which together form the symmetrical weight matrix shown in Fig. 2.51. These weights can be split into two groups as indicated on the diagram. Weights in the first column, and the corresponding identical weights in the first row form the first group. These weights are assigned values reflecting the importance of object i being close to the operator (represented by index 0). For example, assigning a high value to weight w_{io} will tend to result, when e is minimised, in the corresponding distance d_{io} being small. Conversely, giving w_{io} a small value tends to result in the distance d_{io} being large. In this manner, the first design rule is implemented in the evaluation function.

The remaining weights in the matrix form the second group, which are used to express the importance of various objects being close together, or in other words to encode the second design rule. For example, assigning weight w_{ij} a high value implies that objects i and j should be close together. When e is minimised, the resulting distance d_{ij} will tend to be small.

Using the evaluation function (2.43), the layout design problem now takes the form of a mathematical optimisation problem, namely, to determine the locations of the n objects so as to minimise e.

Application to the Workplace Layout Problem. The workplace layout problem is, for several reasons, one which is particularly well suited to the application of a GA. First, the search space contains many local minima which would cause difficulties for conventional hill-climbing optimisation methods, but are easily overcome by a GA. Second, the final solution does not have to be the absolute global minimum for the problem: in general only a reasonably optimal layout is required. Although GAs can theoretically find the global minimum of a problem after a large number of generations, in practice near optimal solutions are usually obtained after relatively few generations. Finally, it is desirable to limit the range of possible object co-ordinates to discrete values to aid in the checking of objects overlapping one another. If objects are assumed to be of unit dimension and object co-ordinates are specified as integers, then two objects i and j will only overlap if:

$$x_i = x_j \text{ and } y_i = y_j \tag{2.45}$$

The integer representation of co-ordinate values is ideally suited to the GA's bit string data storage technique.

To perform the minimisation of the evaluation function (2.43), a GA program was written using the language C. The flowchart for this program is given in Fig. 2.52. The elements of the GA will be discussed in turn.

Initial Population: The evolution process starts with an initial population. Each member of the population comprises a bit string which represents the genes used to carry inherited characteristics in nature. This bit string is employed to encode the co-ordinates x_i, y_i for each object in the layout. Four bits are used for each co-ordinate value allowing values ranging from 0 to 15. Thus to encode the co-ordinates of n objects, the bit string is $8n$ bits long. Figure 2.53 illustrates an example bit string for a layout containing four objects located at co-ordinates (1,2), (4,7), (8,2) and (6,9) respectively.

The initial population is generated randomly, that is, each bit in each bit string is randomly set to either 1 or 0.

	0	1	2	3	4	...	j	...	n	
0	-	W_{01}	W_{02}	W_{03}	W_{04}	...	W_{0j}	...	W_{0n}	Weights representing 1st design rule
1	W_{10}	-	W_{12}	W_{13}	W_{14}	...	W_{1j}	...	W_{1n}	
2	W_{20}	W_{21}	-	W_{23}	W_{24}	...	W_{2j}	...	W_{2n}	
3	W_{30}	W_{31}	W_{32}	-	W_{34}	...	W_{3j}	...	W_{3n}	
...	...	...	...	...	...	...	...	...	...	
i	W_{i0}	W_{i1}	W_{i2}	W_{i3}	W_{i4}	...	W_{ij}	...	W_{in}	Weights representing 2nd design rule
...	...	...	...	...	...	...	...	...	...	
n	W_{no}	W_{n1}	W_{n2}	W_{n3}	W_{n4}	...	W_{nj}	...	-	

Fig. 2.51 Weight matrix representing design rules

Validation: After generating a population, each member must be checked to see that objects in the layout do not overlap. This is done by applying relationship (2.45) to every pair of objects. If overlapping occurs, that member of the population is discarded and a new one generated.

Evaluation: Having obtained a population, a "fitness" value is assigned to each member according to how well it performs as a layout. This value is simply the reciprocal of the evaluation function *e* given in equation (2.43).

Selection, Crossover and Mutation Operators: The operations of these units were explained in Chapter 1.

Swap: The swap operator randomly switches the co-ordinates of two objects. The number of times this operator is applied to members of the population depends on a user-defined swap rate.

Results: Returning to the simple example of the clerical office workstation introduced at the start of this section, the GA method can be used to produce suitable layouts. Here, five objects are to be located, namely, a word processor, a writing area, a printer, a fax machine and a telephone. These objects are assigned indices 1 to 5 respectively, with index 0 being used to represent the operator.

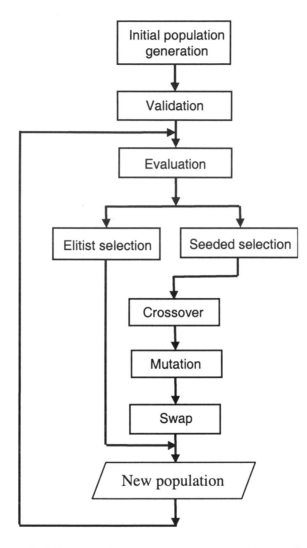

Fig. 2.52 The workplace layout genetic algorithm.

$$\underbrace{1001}_{9}\ \underbrace{0110}_{6}\ \underbrace{0010}_{2}\ \underbrace{1000}_{8}\ \underbrace{0111}_{7}\ \underbrace{0100}_{4}\ \underbrace{0010}_{2}\ \underbrace{0001}_{1}$$

Fig. 2.53 Bit string representation for 4 objects located at co-ordinates (1,2), (4,7), (8,2) and (6,9)

Figure 2.54(a) shows the weight matrix used to encode the design rules and constraint conditions for the problem. Here, weights w_{01} and w_{02} are given the high value of 50 to encourage the GA to place the word processor and writing area closest to the operator. Weight w_{05} is assigned the value 10 to reflect the requirement that the telephone also be reasonably close. Weights W_{15} and W_{25} are assigned values of 10 to encourage the GA to place the telephone close to the word processor and writing area. Weights w_{03} and w_{04} are given the low value 5 since the printer and fax machine need not be placed close to the operator. Finally, weight w_{13} is assigned the value 10 to encourage the GA to place the printer close to the word processor.

Figure 2.54(b) gives layouts resulting from six separate runs of the GA, together with the final evaluation function values. Each run required about 1000 generations to converge to a solution. The best parameters for the GA were found experimentally to be a population size of 100, a scaling window size of six generations, a crossover rate of 0.65 (i.e. crossover applied to 65% of each generation), a mutation rate of 0.1 (i.e. mutation applied to 10% of each generation) and a swap rate of 0.5 (i.e. the swap operator applied to 50% of each generation). The human operator was placed just below the area to which the GA can assign objects (i.e. in row -1).

All of the six layouts generated by the GA are different, despite having similar final evaluation function values. However, in each case the design requirements are met, indicating the existence of multiple solutions to a given layout problem. In all instances, the word processor and the writing area are placed closest to the operator. The telephone is placed close to the word processor and writing area, and the printer is placed next to the word processor.

Figure 2.55 gives the weight matrix and three resulting layouts for a more complex problem of a type typical to an assembly workstation. Here, two groups of five objects are to be positioned. Group *A* consists of objects 1 to 5 and group *B* comprises objects 6 to 10. The objects within each group are to be handled together frequently as represented by the value 10 assigned to weights w_{12} and w_{13} etc. Objects in group *B* are handled more frequently than *A*, represented by the value 100 assigned to weights w_{06} to w_{010} as opposed to the value 50 assigned to weights w_{01} to w_{05}. In all of the layouts produced by the GA, the objects within each group are located adjacent to each other. In the first two cases, group *A* objects are located closest to the operator.

	0	1	2	3	4	5
0	-	50	50	5	5	10
1	50	-	0	10	0	10
2	50	0	-	0	0	10
3	5	10	0	-	0	0
4	5	0	0	0	-	0
5	10	10	10	0	0	-

Index	Object
0	Operator
1	Word Processor
2	Writing Area
3	Printer
4	Fax Machine
5	Telephone

(a) Weight Matrix

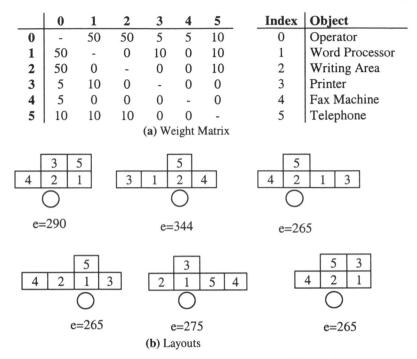

(b) Layouts

Fig. 2.54 Application of the GA to the layout of a clerical office workstation

	0	1	2	3	4	5	6	7	8	9	10
0	-	50	50	50	50	50	100	100	100	100	100
1	50	-	10	10	10	10	0	0	0	0	0
2	50	10	-	10	10	10	0	0	0	0	0
3	50	10	10	-	10	10	0	0	0	0	0
4	50	10	10	10	-	10	0	0	0	0	0
5	50	10	10	10	10	-	0	0	0	0	0
6	100	0	0	0	0	0	-	10	10	10	10
7	100	0	0	0	0	0	10	-	10	10	10
8	100	0	0	0	0	0	10	10	-	10	10
9	100	0	0	0	0	0	10	10	10	-	10
10	100	0	0	0	0	0	10	10	10	10	-

Index	Object
0	Operator
1	⎫
2	
3	Group 1
4	
5	⎭
6	⎫
7	
8	Group 2
9	
10	⎭

(a) Weight Matrix

e = 1862   e = 1809   e = 200

(b) Layouts

Fig. 2.55 Application of the GA to a two-group layout problem.

Figure 2.56 gives the weight matrix and three resulting layouts for a second group layout problem. This time, four groups of objects are to be located. Group A consists of objects 1 to 3, group B of objects 4 to 6, group C of objects 7 and 8 and group D of objects 9 and 10. As before, the objects within each group are to be handled together frequently, as represented by the value 10 assigned to weights w_{12} and w_{13} etc. This time however, each group is handled equally often, and the value 60 is assigned to all the weights w_{01} to w_{010}. Also, the operator is located within the area to which the GA can assign objects. In each of the layouts generated by the GA, objects within the four groups are located in close proximity, around the operator.

	0	**1**	**2**	**3**	**4**	**5**	**6**	**7**	**8**	**9**	**10**
0	-	60	60	60	60	60	60	60	60	60	60
1	60	-	10	10	0	0	0	0	0	0	0
2	60	10	-	10	0	0	0	0	0	0	0
3	60	10	10	-	0	0	0	0	0	0	0
4	60	0	0	10	-	10	10	0	0	0	0
5	60	0	0	0	10	-	10	0	0	0	0
6	60	0	0	0	10	10	-	0	0	0	0
7	60	0	0	0	0	0	0	-	10	0	0
8	60	0	0	0	0	0	0	10	-	0	0
9	60	0	0	0	0	0	0	0	0	-	10
10	60	0	0	0	0	0	0	0	0	10	-

Index	Object
0	Operator
1	Group 1
2	
3	
4	Group 2
5	
6	
7	Group 3
8	
9	Group 4
10	

(a) Weight Matrix

10	6	5
9	O	4
3	2	7
	1	8

e = 1491

	3	
1	7	8
2	O	4
9	10	6

e = 1492

1	10	9	
2	O	4	6
3	7	5	
	8		

e = 1462

(b) Layouts

Fig. 2.56 Application of the GA to a four-group layout problem.

Finally, Fig. 2.57 gives the weight matrix and three resulting layouts for the last layout problem considered in this section. Here, there are no groupings amongst objects and the weights w_{12} and w_{13} etc. are accordingly set to zero. However, objects 1 to 10 are to be handled with decreasing frequency, and hence weights w_{01} and w_{010} are assigned decreasing values 100, 90, 80 etc. The operator is once more located outside of the area to which the GA can assign objects (i.e. in row -1). In each of the layouts generated by the GA, objects are located in an order of distance from the operator which reflects the handling frequency.

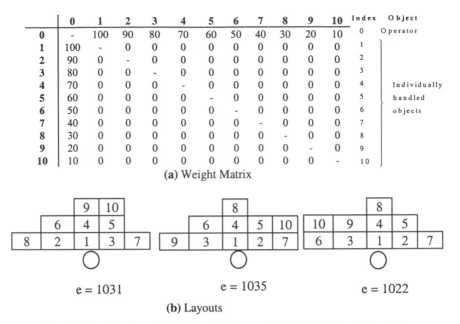

(a) Weight Matrix

e = 1031

e = 1035

e = 1022

(b) Layouts

Fig. 2.57 Application of the GA to the layout of individually handled objects.

2.3 Summary

This chapter contains two main sections. In the first, four new versions of genetic algorithms are described. The first of these models has a novel reproduction operator, which overcomes the drawbacks of roulette wheel selection. This new genetic algorithm is a version of a micro GA. The second version is a simple genetic algorithm that can avoid premature convergence by simultaneously searching different areas of the solution space. The proposed GA is eminently suitable for parallelisation, which would increase the speed of operation. In the third version, four techniques are employed to help a GA to find multiple, alternative solutions for a problem. The main advantage of these four techniques is that they all encourage genetic variety, and hence the production of different solutions, without disrupting the gene population. It is well known that the mutation operation is most important to the performance of a GA when searching a highly complex solution space as it helps the search process to escape from local minima and find better solutions. The fourth GA version employs three simple strategies for automatically adapting the mutation rate in a genetic algorithm.

The second section of this chapter describes different applications of GAs to problems from different engineering areas. The first application is the design of static fuzzy logic controllers. A standard GA is employed and the performance of the FLCs obtained compared with that of a classical PID controller with parameters optimised in terms of the transient response. The second application uses a GA to train recurrent neural networks for the identification of dynamic systems. The neural networks employed are the Elman network, the Jordan network and their modified versions. Identification results are given for linear and non-linear systems. In the third application, a simple GA is used for the design of an adaptive fuzzy logic controller and the simulation results were produced for linear and nonlinear dynamic systems. The fourth application example is that of gearbox design. The last application is the automatic generation of workplace layouts. Results are given for a number of layout problems, proving the effectiveness of the GA method.

References

Baluja, S. (1993) Structure and performance of fine-grain parallelism in genetic search, *5th Int. Conf. on Genetic Algorithms and their Applications*, Morgan Kaufmann, San Mateo, CA, pp.155-162.

Brown, K. M. and Gearhart, W. B. (1971) Deflation techniques for the calculation of further solutions of a non-linear system. *Numer. Math.*, Germany, Vol.16, No.4, pp.334-342.

Chalmers, D.J., (1990) The evolution of learning: an experiment in genetic connectionism, In: Touretzky, D.S., Elman, J.L. and Hinton, G.E., (eds), *Proc. Connectionist Models Summer School*, pp.81-90, Morgan Kaufmann, San Mateo, CA.

Chen, C. (1984) *Linear System Theory and Design*, Holt, Rinehart and Winston, NewYork.

Das B., Grady R.M. (1983) Industrial workplace layout and engineering anthropology. In Kvalseth T.O (ed.) *Ergonomics of Workstation Design*, Butterworth, London, pp.103-125.

Davis, L. (1989) Adapting operator probabilities in genetic algorithms, *Proc. 3rd Int. Conf. on Genetic Algorithms and Their Applications*, George Mason University, pp.61-69.

Davis, L. (1985) Job shop scheduling with genetic algorithms. *Proc. Int. Conf. on Genetic Algorithms and their Applications*, Pittsburgh, Carnegie-Mellon University, pp.136-140.

Davis, L. (1991) *Handbook of Genetic Algorithms*, Van Nostrand Reinhold, New York, NY.

De Jong, K.A. (1975) *An Analysis of the Behavior of a Class of Genetic Adaptive Systems*, PhD Dissertation, University of Michigan, Ann Arbor, Michigan.

Elman, J.L. (1990) Finding structure in time, *Cognitive Science*, Vol.14, pp.179-211.

Fogarty, T.C. (1989) Varying the probability of mutation in the genetic algorithm, *Proc. 3rd Int. Conf. on Genetic Algorithms and their Applications*, George Mason University, pp.104-109.

Fourman M.P. (1985) Compaction of symbolic layout using genetic algorithms. *Proc. Int. Conf. on Genetic Algorithms and their Applications*, Pittsburgh, Carnegie-Mellon University, pp 141-153

Goldberg, D. E. and Richardson, J. (1987) Genetic algorithms with sharing for multimodal optimization. *2nd International Conference on Genetic Algorithms and Their Applications,* Lawrence Erlbaum Associates, Hillsdale, NJ, pp.41-49.

Goldberg, D.E. (1985) *Optimal Initial Population Size for Binary-Coded Genetic Algorithms*, TCGA Report Number 8510001, Department of Engineering Mechanics, University of Alabama, Alabama.

Goldberg, D.E. (1989) *Genetic Algorithms in Search, Optimization, and Machine Learning*, Addison-Wesley, Reading, MA.

Goldberg, D.E. (1989) Sizing populations for serial and parallel genetic algorithms, *Proc. 3rd Int. Conf. on Genetic Algorithms*, Fairfax, VA, pp.70-79.

Gordan, V.S. and Whitely, D. (1993) Serial and parallel genetic algorithms as function optimizers, *5ʰ Int. Conf. on Genetic Algorithms and their Applications*, Morgan Kaufmann, San Mateo, CA, pp.177-183.

Grefenstette, J.J. (1986) Optimization of control parameters for genetic algorithms, *IEEE Trans. on Systems Man and Cybernetics*, Vol.SMC-16, No.1, pp.122-128.

Gupta, M.M., Kiszka, J.B. and Trojan, G.M. (1986) Multi variable structure of fuzzy control systems, *IEEE Trans. on Systems, Man and Cybernetics*, Vol.SMC-16, No.5, pp.638-656.

Harp, S.A. and Samad T. (1991) Genetic synthesis of neural network architecture, In: Davis, L. (ed.), *Handbook of Genetic Algorithms*, pp.203-221, Van Nostrand Reinhold, New York, NY.

Holland, J.H. (1975) *Adaptation in Natural and Artificial Systems*, University of Michigan Press, Ann Arbor, MI.

Jin, G. (1995) *Intelligent Fuzzy Logic Control of Processes with Time Delays*, PhD Thesis, School of Engineering, Cardiff University, UK.

Jordan, M.I. (1986) Attractor dynamics and parallelism in a connectionist sequential machine, *Proc. 8ʰ Annual Conf. of the Cognitive Science Society*, Amherst, MA, pp.531-546.

Krishnakumar, K. (1989) Micro genetic algorithms for stationary and non-stationary function optimization, *SPIE Vol.1196, Intelligent Control and Adaptive Systems*, pp.289-296.

Ku, C.C. and Lee, K.Y. (1995) Diagonal recurrent neural networks for dynamic systems control, *IEEE Trans. on Neural Networks*, Vol.6, No.1, pp.144-156.

Lee, C.C. (1990a) Fuzzy logic in control systems: Fuzzy logic controller, Part I, *IEEE Trans. on Systems, Man and Cybernetics*, Vol.20, No.2, pp.404-418.

Lee, C.C. (1990b) Fuzzy logic in control systems: Fuzzy logic controller, Part II, *IEEE Trans. on Systems, Man, and Cybernetics*, Vol.20, No.2, pp.419-435.

Linkens, D.A. and Hanain, S.B. (1991) Self- organising fuzzy logic control and application to muscle relaxant anaesthesia, *IEE Proceedings, Part D,* Vol.138, No.3, pp.274-284.

Manderick, B. and Spiessens, P. (1989) Fine-grained parallel genetic algorithms, *3rd Int. Conf. on Genetic Algorithms and their Applications*, Morgan Kaufmann, San Mateo, CA, pp.428-433.

Mansfield, R.A. (1990) *Genetic Algorithms*. Master's Thesis, School of Engineering, Cardiff University, UK.

Maruyama, T., Hirose, T. and Konagaya, A. (1991) A fine-grained parallel genetic algorithm for distributed parallel systems, *4th Int. Conf. on Genetic Algorithms and their Applications*, Morgan Kaufmann, San Mateo, CA, pp.184-190.

McCormick E.J. and Sanders M.S. (1982) *Human Factors in Engineering and Design*, McGraw-Hill, New York, NY.

Moore, C.G. and Harris, C.J. (1992) Indirect adaptive fuzzy control, *Int. J. Control*, Vol.56, No.2, pp.441-468.

Muhlenbein, H. (1989) Parallel genetic algorithms, population genetics and combinatorial optimization, *3rd Int. Conf. on Genetic Algorithms and Their Applications*, Morgan Kaufmann, San Mateo, CA, pp.416-421.

Muselli, M. (1992) Global optimization of functions with the interval genetic algorithm, *Complex Systems*, Vol.6, pp.193-212.

Narendra, K.S. and Annaswamy, A.M. (1989) *Stable Adaptive Systems*, Prentice-Hall, Englewood Cliffs, NJ.

Oh, J. (1993) *Identification and Control of Dynamic Systems Using Neural Networks*, PhD. Thesis, School of Engineering, Cardiff University, UK.

Peng, X.T. (1990) Generating rules for fuzzy logic controllers by functions, *Fuzzy Sets and Systems*, Vol.36, pp.83-89.

Peng, X.T., Liu, S.M., Yamakawa, T., Wang, P. and Liu, X. (1988) Self-regulating PID controllers and its applications to a temperature controlling process, *Fuzzy Computing*, M.M. Gupta and T. Yamakawa (eds.), Elsevier, Amsterdam, pp.355-364.

Pham, D.T. and Jin, G. (1996) A hybrid genetic algorithm, *3rd World Congress on Expert Systems*, Seoul, Korea, Vol. II, pp.748-757.

Pham, D.T. and Karaboga, D. (1991a) A new method to obtain the relation matrix of fuzzy logic controllers, *Proc. 6ᵗʰ Int. Conf. on Artificial Intelligence in Engineering*, Oxford, pp.567-581.

Pham, D.T. and Karaboga, D. (1991b) Optimum design of fuzzy logic controllers using genetic algorithms, *J. of Systems Engineering*, Vol.1, No.2, pp.114-118.

Pham, D.T. and Karaboga D. (1994a) *Training Elman and Jordan networks for system identification using genetic algorithms*, Technical Report, Intelligent Systems Laboratory, School of Engineering, Cardiff University, UK.

Pham, D.T. and Karaboga D. (1994b) Design of an adaptive fuzzy logic controller, *IEEE Int. Conf. On Systems, Man and Cybernetics (IEEE-SMC'94)*, Vol.1, pp.437-442.

Pham, D.T. and Karaboga, D. (1997) Genetic algorithms with variable muattion rates: application to fuzzy logic controller design, *Proc. Instn. Mech. Engrs*. Vol.211, Part1, pp.157-167.

Pham, D.T. and Liu, X. (1992) Dynamic system identification using partially recurrent neural networks, *J. of Systems Engineering*, Vol.2, No.2, pp.90-97.

Pham, D.T. and Liu, X. (1995) *Neural Networks for Identification, Prediction and Control* (3ʳᵈ printing), Springer-Verlag, London.

Pham, D.T. and Liu, X. (1996) Training of Elman networks and dynamic system modelling, *Int. J. of Systems Science*, Vol.27, No.2, pp.221-226.

Pham, D.T. and Oh, S.J. (1992) A recurrent backpropagation neural network for dynamic system identification, *J. of Systems Engineering*, Vol.2, No.4, pp.213-223.

Pham D.T. and Onder, H.H. (1992) A knowledge-based system for optimising workplace layouts using a genetic algorithm, *Ergonomics*, Vol.35, No.12, pp.1479-1487.

Pham, D.T., Onder, H.H. and Channon, P.H. (1996) Ergonomic workplace layout using genetic algorithms, *J. of Systems Engineering,* Vol.6, pp.119-125.

Pham, D.T. and Yang, Y. (1993) Optimization of multi-modal discrete functions using genetic algorithms, *Proc. Instn. Mech. Engrs.*, Part D, Vol.207, pp.53-59.

Potts, J.C., Giddens, T.D. and Yadav, S.B. (1994) The development and evaluation of an improved genetic algorithm based on migration and artificial selection, *IEEE Trans. Syst., Man and Cybernetics*, Vol.24, No.1, pp.73-86.

Powell, D.J., Skolnick, M.M. and Tong, S.S. (1990) EnGENous: a unified approach to design optimization, *Proc. 5th Int. Conf. on Applications of Artificial Intelligence in Engineering*, July, Computational Mechanics Publishers and Springer-Verlag, Boston, MA, pp.137-157.

Procyk, T.J. and Mamdani, E.H. (1979) A linguistic self-organizing process controller, *Automatica*, Vol.15, pp.15-30.

Rumelhart, D.E. and McClelland, J.L., (1986) *Parallel Distributed Processing : Explorations in the Micro-Structure of Cognition*, Vol.1, MIT Press, Cambridge, MA.

Schaffer, J.D., Caruana, R.A., Eshelman, L.J. and Das, R. (1989) A study of control parameters affecting on-line performance of genetic algorithms for function optimisation, *Proc. 3rd Int. Conf. on Genetic Algorithms and their Applications*, George Mason University, pp.51-61.

Shao, S. (1988) Fuzzy self-organizing controller and its application for dynamic process, *Fuzzy Sets and Systems*, Vol.26, pp.151-164.

Starkweather, T., Whitely, D. and Mathias, K. (1991) *Optimization Using Distributed Genetic Algorithms*, Lecture notes in Computer Science, Vol.496, Springer-Verlag, Heidelberg.

Sugiyama, K. (1986) *Analysis and Synthesis of the Rule-Based Self-Organising Controller*, Ph.D. Thesis, Queen Mary College, University of London.

Whitely, D. and Hanson, T. (1989) Optimising neural networks using faster, more accurate genetic search, *Proc. 3rd Int. Conf. on Genetic Algorithms and their Applications*, George Mason University, pp.370-374.

Whitely, D. and Hanson, T. (1989) Optimising neural networks using faster, more accurate genetic search, In: Schaffer, J.D. (ed.), *Proc. 3rd Int. Conf. on Genetic Algorithms and their Applications*, Morgan Kaufmann, San Mateo, CA, pp.370-374.

Yamazaki, T. (1982) *An Improved Algorithm for a Self Organising Controller*, Ph.D. Thesis, Queen Mary College, University of London.

Chapter 3

Tabu Search

This chapter consists of five main sections describing the application of the tabu search algorithm to several engineering problems. The first two sections discuss the use of tabu search to optimise the effective side-length expression for the resonant frequency of triangular microstrip antennas [Karaboga *et al.*, 1997a] and to obtain a simple formula for the radiation efficiency of rectangular microstrip antennas. The third section describes the application of tabu search for training recurrent neural networks [Karaboga and Kalinli, 1996a; 1997]. The fourth section considers the optimal design of digital Finite-Impulse-Response (FIR) filters using tabu search [Karaboga *et al.*, 1997b]. The last section presents the application of tabu search to tuning Proportional-Integral-Derivative (PID) controller parameters [Karaboga and Kalinli, 1996c].

3.1 Optimising the Effective Side-Length Expression for the Resonant Frequency of a Triangular Microstrip Antenna

In recent years, microstrip antennas have aroused great interest in both theoretical research and engineering applications. This is due to their low profile, light weight, conformal structure, low cost, reliability, ease of fabrication and integration with solid-state devices [Bahl and Bhartia, 1980, Carver and Mink, 1981; James *et al.*, 1981; James and Hall, 1989; Lo *et al.*, 1989; Zürcher and Gardiol, 1995].

The majority of studies proposed in this area have concentrated on rectangular and circular microstrip antennas. It is known that the triangular patch antenna has radiation properties similar to the rectangular antenna but with the advantage of being smaller. Triangular microstrip antennas are of particular interest for the design of periodic arrays because triangular radiating elements can be arranged so that the designer can reduce significantly the coupling between adjacent elements of the array. This simplifies array design significantly. In triangular microstrip antenna design, it is important to determine the resonant frequencies of the antenna accurately because microstrip antennas have narrow bandwidths and can only operate effectively in the vicinity of the resonant frequency. As such, a theory to help ascertain the resonant frequency is helpful in antenna design.

The resonant frequency of such antennas is a function of the side length of the patch, the permittivity of the substrate and its thickness. A number of methods are available to determine the resonant frequency for an equilateral triangular microstrip patch antenna, as this is one of the most popular and convenient shapes [Bahl and Bhartia, 1980; Chen *et al.*, 1992; Dahele and Lee, 1987; Gang, 1989; Garg and Long, 1988; Güney, 1993b; Güney, 1994a; Helszajn and James, 1978; Kumprasert and Kiranon, 1994; Singh and Yadava, 1991]. Experimental resonant frequency results of this antenna have been reported in [Chen *et al.*, 1992; Dahele and Lee, 1987]. The theoretical resonant frequency values presented in the literature, however, do not correspond well with the experimental results. For this reason, an effective new side-length expression is presented here for an equilateral triangular patch antenna. The resonant frequencies of this antenna are then obtained by using this expression and the relative dielectric constant of the substrate.

In this work, first, a model is chosen for the effective side-length expression. Second, a modified tabu search algorithm is used to obtain the unknown coefficient values of the expression. The tabu search algorithm used here employs an adaptive neighbour production mechanism. Therefore, this algorithm is different from the tabu search algorithms in the literature [Glover, 1989; Glover, 1990]. This algorithm can also have other applications in computer-aided design (CAD) of microstrip antennas and microwave integrated circuits. The results obtained demonstrate the versatility, robustness and computational efficiency of the algorithm.

The theoretical resonant frequency results obtained using the new side-length expression correspond well with the experimental results [Chen *et al.*, 1992; Dahele and Lee, 1987]. The new side-length expression is also simple and useful to antenna engineers for accurately predicting the resonant frequencies of equilateral triangular microstrip patch antennas. Güney [1993b; 1994b, c] also proposed very simple expressions for calculating accurately the resonant frequencies of rectangular and circular microstrip antennas.

Most of the previous theoretical resonant frequency results for triangular microstrip antennas were compared only with the experimental results reported by Dahele and Lee [1987]. Here, the theoretical results obtained using the formulas available in the literature are compared with the experimental results reported by Dahele and Lee [1987] and also Chen *et al.* [1992].

3.1.1 Formulation

For a triangular microstrip antenna, the resonant frequencies obtained from the cavity model with perfect magnetic walls are given by the formula [Helszajn and James, 1978]:

$$f_{mn} = \frac{2c}{3a(\varepsilon_r)^{1/2}} \left[m^2 + mn + n^2 \right]^{1/2} \tag{3.1}$$

where c is the velocity of electromagnetic waves in free space, ε_r is the relative dielectric constant of the substrate, subscript mn refers to TM_{mn} modes, where TM stands for transverse magnetic and a is the length of a side of the triangle, as shown in Figure 3.1.

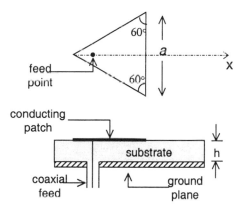

Fig. 3.1 Geometry of equilateral triangular microstrip antenna

Equation (3.1) is based on the assumption of a perfect magnetic wall and neglects the fringing fields at the open-end edge of the microstrip patch. To account for these fringing fields, there have been a number of suggestions [Bahl and Bhartia, 1980; Chen *et al.*, 1992; Dahele and Lee, 1987; Gang, 1989; Garg and Long, 1988; Güney, 1993a, b, c; Güney, 1994a; Helszajn and James, 1978; Kumprasert and Kiranon, 1994; Singh and Yadava, 1991]. The most common suggestion is that the side length a in equation (3.1) be replaced by an effective value a_{eff}. This suggestion is adopted here.

The effective side-length a_{eff}, which is slightly larger than the physical side length a, takes into account the influence of the fringing field at the edges and the dielectric inhomogenity of the triangular microstrip patch antenna. The effective side length of a triangular microstrip antenna is determined by the relative dielectric constant of the substrate ε_r, the physical side-length a and the thickness of the substrate h.

The problem in the literature is that as simple an expression as possible for the effective side length should be derived but the theoretical results obtained by using the expression must agree well with the experimental results. In this work, a new technique based on tabu search is presented for solving this problem. First, a model for the effective side-length expression is chosen. Then, a modified tabu search algorithm is employed to determine the unknown coefficients of the model.

To find a suitable model for the effective side-length expression, many experiments were carried out and the following model, dependent on the values of ε_r, a and h that produce good results, was chosen:

$$a_{eff} = a + h\left(\alpha_1 + \frac{\alpha_2}{\varepsilon_r^{\alpha_3}}\right) \qquad (3.2)$$

where the unknown coefficients α_1, α_2 and α_3 are to be determined by a modified tabu search algorithm. It is evident from equation (3.2) that the effective side length a_{eff} is larger than the physical side length a provided that the term inside the brackets is greater than zero. A way of satisfying this condition is for α_1, α_2 and α_3 all to be positive.

The tabu search algorithm utilised in this work employs an adaptive mechanism for producing neighbours. The neighbours of a present solution are created by the following procedure.

If $a_{eff}(t) = (\alpha_1 \, \alpha_2 \, \alpha_3)$ is the solution vector at the tth iteration, two neighbours of this solution, whose element α_k is not on the tabu list, are produced by:

$$a_{eff}\left(\overline{n}_1, \overline{n}_2\right) = \begin{cases} \alpha_k + \Delta(t) & \textit{for odd neighbours} \\ \alpha_k - \Delta(t) & \textit{for even neighbours} \end{cases} \qquad (3.3)$$

$$a_{eff}\left(n_1, n_2\right) = remain\left(a_{eff}\left(\overline{n}_1, \overline{n}_2\right), \alpha_{max}\right)$$

with

$$\Delta(t) = k_1\left[\frac{LatestImprovementIteration}{Iteration^{k_2} + LatestImprovementIteration}\right]^{k_3} \qquad (3.4)$$

where   *Iteration*   stands   for   the   current   iteration   number   and

LatestImprovementIteration is the iteration number at which the latest improvement was obtained. The value of α_{max}, which is larger than zero for each coefficient, is determined by the designer. The index t in $\Delta(t)$ represents the iteration number. The *remain* function keeps the elements of the solution within the desired range. k_1 in equation (3.4) determines the magnitude of $\Delta(t)$ while k_2 and k_3 control the change of $\Delta(t)$ with t. Suitable values for k_1, k_2 and k_3 are determined experimentally. Here, the values are taken as 10, 2 and 2, respectively.

The tabu restrictions employed are based on the *recency* and *frequency memory* criteria. If an element of the solution vector does not satisfy one of the following tabu restrictions, then it is accepted as being tabu:

$$\text{Tabu restrictions} = \begin{cases} recency > rN \\ \\ frequency < f.Avfreq \end{cases} \tag{3.5}$$

where N is the length of the binary string representing the solution vector, *Avfreq* stands for the average frequency of change of bits in the string and r and f are recency and frequency factors for the tabu restrictions, respectively.

In equation (3.5), the value of N is 24 since each parameter is represented by 8 bits, r equals 0.5 and f equals 2.

To select the new solution from the non-tabu neighbours, the performances of the neighbours are computed and the neighbour producing the highest improvement with respect to the present solution is chosen as the next solution. If none of the neighbours are better than the present solution, then the neighbours are again evaluated using the following formula:

$$evaluation(i) = a.improvement(i) + b.recency(i) - c.frequency(i) \tag{3.6}$$

where a, b and c are the improvement, recency and frequency factors used for the evaluation and are equal to 4, 2 and 1, respectively. In equation (3.6), *improvement* is the difference between the performance of the present solution and that of the ith neighbour. The performance of a neighbour can be computed using various formulas. Here, the following formula is employed:

$$P(i) = A - \sum_{j=1}^{M} \left| f(j)_{me} - f(j)_{ca} \right| \tag{3.7}$$

where A is a positive constant selected to be large enough so that $P(i)$ is positive for all possible solutions. Here, A is taken as 1000. f_{me} represents the measured resonant frequency values and f_{ca} the calculated resonant frequency values using the effective side-length expression constructed by the tabu search algorithm. The measured data used for the optimisation and evaluation process were obtained from previous work and are shown in the row 1 of Tables 3.1-3.3. The $TM_{30} f_{me}$ entries in these tables are used for the evaluation process to demonstrate the accuracy of the model while the remaining 12 f_{me} entries (TM_{10}, TM_{11}, TM_{20} and TM_{21}; $M=12$ in Eq. (3.7)) are used for the optimisation process. Only 3 measured (TM_{30}) data values are used for the evaluation process because of the limited measured data available in the literature.

The modified tabu search algorithm described above found the optimal coefficient values for the model given in equation (3.2) as:

$$\alpha_1 = 0.1 \qquad \alpha_2 = 8 \qquad \alpha_3 = 2 \tag{3.8}$$

Then, the following effective side-length expression a_{eff} is obtained by substituting these coefficient values into equation (3.2):

$$a_{eff} = a + h\left(0.1 + \frac{8}{\varepsilon_r^2}\right) \tag{3.9}$$

Table 3.1 Measured and calculated resonant frequencies of the first five modes of an equilateral triangular microstrip antenna with $a = 10$cm, $\varepsilon_r = 2.32$ and $h = 0.159$cm

	Modes				
Methods	TM_{10}	TM_{11}	TM_{20}	TM_{21}	TM_{30}
f_{me} [Measured, Dahale and Lee,1987]	1280	2242	2550	3400	3824
f_{pm} (Proposed method)	1281	2218	2562	3389	3842
f_{bb} [Bahl and Bhartia,1980]	1413	2447	2826	3738	4239
f_{hj} [Helszajn and James,1978]	1299	2251	2599	3438	3898
f_{gl} [Garg and Long,1988]	1273	2206	2547	3369	3820
f_{ga} [Gang,1989]	1340	2320	2679	3544	4019
f_{sd} [Singh and Yadava,1991]	1273	2206	2547	3369	3820
f_{cl1} [Chen et al.,1992] (Moment method)	1288	2259	2610	3454	3874
f_{cl2} [Chen et al.,1992] (Curve-fitting method)	1296	2244	2591	3428	3887
f_{gu1} [Güney,1993b]	1280	2217	2560	3387	3840
f_{gu2} [Güney,1994a]	1280	2218	2561	3387	3841
f_{kk} [Kumprasert and Kiranon,1994]	1289	2233	2579	3411	3868

Table 3.2 Measured and calculated resonant frequencies of the first five modes of an equilateral triangular microstrip antenna with $a = 8.7$cm, $\varepsilon_r = 2.32$ and $h = 0.078$cm

Methods	Modes				
	TM_{10}	TM_{11}	TM_{20}	TM_{21}	TM_{30}
f_{me} [Measured, Chen et al.,1992]	1489	2596	2969	3968	4443
f_{pm} (Proposed method)	1488	2577	2976	3937	4464
f_{bb} [Bahl and Bhartia,1980]	1627	2818	3254	4304	4880
f_{hj} [Helszajn and James,1978]	1500	2599	3001	3970	4501
f_{gl} [Garg and Long,1988]	1480	2564	2961	3917	4441
f_{ga} [Gang,1989]	1532	2654	3065	4054	4597
f_{sd} [Singh and Yadava,1991]	1480	2564	2961	3917	4441
f_{cl1} [Chen et al.,1992] (Moment method)	1498	2608	2990	3977	4480
f_{cl2} [Chen et al.,1992] (Curve-fitting method)	1498	2595	2996	3963	4494
f_{gu1} [Güney,1993b]	1486	2573	2971	3931	4457
f_{gu2} [Güney,1994a]	1481	2565	2962	3918	4443
f_{kk} [Kumprasert and Kiranon,1994]	1493	2585	2985	3949	4478

Table 3.3 Measured and calculated resonant frequencies of the first five modes of an equilateral triangular microstrip antenna with $a = 4.1$ cm, $\varepsilon_r = 10.5$ and $h = 0.07$ cm

Methods	Modes				
	TM_{10}	TM_{11}	TM_{20}	TM_{21}	TM_{30}
f_{me} [Measured, Chen et al.,1992]	1519	2637	2995	3973	4439
f_{pm} (Proposed method)	1501	2600	3002	3971	4503
f_{bb} [Bahl and Bhartia,1980]	1725	2988	3450	4564	5175
f_{hj} [Helszajn and James,1978]	1498	2594	2995	3962	4493
f_{gl} [Garg and Long,1988]	1494	2588	2989	3954	4483
f_{ga} [Gang,1989]	1577	2731	3153	4172	4730
f_{sd} [Singh and Yadava,1991]	1494	2588	2989	3954	4483
f_{cl1} [Chen et al.,1992] (Moment method)	1522	2654	3025	4038	4518
f_{cl2} [Chen et al.,1992]	1509	2614	3018	3993	4528
f_{gu1} [Güney,1993b]	1511	2617	3021	3997	4532
f_{gu2} [Güney,1994a]	1541	2669	3082	4077	4623
f_{kk} [Kumprasert and Kiranon,1994]	1490	2581	2980	3942	4470

3.1.2 Results and Discussion

To determine the most appropriate suggestion given in the literature, the computed values of the resonant frequencies for the first five modes of different equilateral triangular patch antennas were compared with the theoretical and experimental results reported by other researchers, which are given in Tables 3.1-3.3. f_{pm} represents the value obtained by the new method. f_{me} represents the value obtained experimentally by Dahele and Lee [1987] (Table 3.1) and by Chen *et al.* [1992] (Tables 3.2 and 3.3). f_{bb} , f_{hj} , f_{gl} , f_{ga} , f_{sd} , f_{cl1} , f_{cl2} , f_{gu1} , f_{kk} and f_{gu2} represent, respectively the values calculated by Bahl and Bhartia [1980], Helszajn and James [1978], Garg and Long [1988], Gang [1989], Singh and Yadava [1991], Chen *et al.* [1992](moment method), Chen *et al.* [1992](curve-fitting formula), Güney [1993b], Kumprasert and Kiranon [1994] and Güney [1994a]. The total absolute errors between the theoretical and experimental results in Tables 3.1-3.3, for each model, are listed in Table 3.4. The theoretical results predicted by Garg and Long [1988] and Singh and Yadava [1991] are the same because their analytical formulas are also the same.

It can be seen from Tables 3.1-3.4 that the theoretical resonant frequency results of previous researchers generally do not correspond well with the experimental results. Therefore, the data sets obtained from existing theories are not used here. The measured data set only is employed for the optimisation process.

The proposed side-length expression a_{eff} is accurate in the range $2.3 < \varepsilon_r < 10.6$ and $0.005 < (h/\lambda_d) < 0.034$, where λ_d is the wavelength in the substrate. From Tables 3.1-3.4, the results calculated by the new expression are generally better than those predicted in previous work. The very good correlation between the measured values and the computed resonant frequency values supports the validity of the new side-length expression obtained using the modified tabu search algorithm, even though the data set is limited. The theoretical results reported by Garg and Long [1988], Kumprasert and Kiranon [1994] and Güney [1993b] also agree quite well with the experimental results. However, the new formula is simpler than those adopted by these researchers.

Since the formula presented here has a good accuracy and is simple, it can be very useful for the development of fast CAD algorithms. It is also very useful to antenna engineers. Using this formula, one can calculate accurately by hand, the resonant frequency of triangular patch antennas, without possessing any background knowledge of microstrip antennas.

It should be emphasised again that better and more robust results might be obtained from the modified tabu search algorithm if more experimental data were supplied for the optimisation process.

Table 3.4 Total absolute errors between the measured and calculated resonant frequencies

Method	Error
f_{pm} (Proposed method)	273
f_{bb} [Bahl and Bhartia,1980]	5124
f_{hj} [Helszajn and James,1978]	424
f_{gl} [Garg and Long,1988]	326
f_{ga} [Gang,1989]	1843
f_{sd} [Singh and Yadava,1991]	326
f_{cl2} [Chen et al.,1992]	408
f_{gu1} [Güney,1993b]	314
f_{gu2} [Güney,1994a]	590
f_{kk} [Kumprasert and Kiranon,1994]	349

3.2 Obtaining a Simple Formula for the Radiation Efficiency of a Resonant Rectangular Microstrip Antenna

Efficient usage of microstrip antennas requires knowledge of radiation efficiency, which relates the power radiated in space waves to the total radiated power (including surface waves). The surface waves exist due to the contact between air and the dielectric substrate that separates the radiating element from the ground plane. For infinitely long and lossless structures, the surface wave propagates and attenuates in directions parallel and perpendicular to the air/dielectric interface, respectively. Surface wave power launched in an infinitely-wide substrate would not contribute to the main beam radiation and so can be treated as a loss mechanism. In general, if the thickness of the substrate on which the antenna is etched is very small compared to the wavelength of interest, the power propagated via the surface wave modes is negligible so that the effects of the substrate on the efficiency may be ignored. However, for an antenna radiating at the resonance of higher-order modes, the radiation efficiency decreases as significant power propagates via surface wave modes. Moreover, unwanted radiation results when the surface wave encounters a discontinuity (e.g. the edge of the substrate). As a surface wave reaches the antenna's edge, it is scattered, producing both a reflected surface wave and a radiated wave. The presence of secondary sources of radiation on the dielectric edges has proved troublesome in practice, as they contribute to secondary lobes and to cross-polarised radiation. A large surface wave excitation also causes an undesirable energy coupling between elements of an array or between adjacent arrays.

The surface wave effect has been identified by a number of investigators [Bhattacharyya and Garg, 1986; Güney, 1993a, b, c; Güney, 1995; Jackson and Alexopoulos, 1991; James and Henderson, 1979; Lo *et al.*, 1979; Mosig and Gardiol, 1983; Mosig and Gardiol, 1985; Nauwelaers and Van De Capelle, 1989; Perlmutter *et al.*, 1985; Pozar, 1983; Rana and Alexopoulos, 1981; Roudot *et al.*, 1985; Uzunoglu *et al.*, 1979; Wood, 1981]. Uzunoglu *et al.* [1979] determined the radiation efficiency of dipoles by employing a dyadic Green function for a Hertzian dipole printed on a grounded substrate together with an assumed current distribution. James and Henderson [1979] estimated that surface wave excitation is not important if $h/\lambda_0 < 0.09$ for $\varepsilon_r \cong 2.3$ and $h/\lambda_0 < 0.03$ for $\varepsilon_r \cong 10$, where h is the thickness of the dielectric substrate and λ_0 is the free-space wavelength. The criterion presented by Wood [1981] is more quantitative: $h/\lambda_0 < 0.07$ for $\varepsilon_r = 2.3$ and $h/\lambda_0 < 0.023$ for $\varepsilon_r = 10$ if the antenna is to launch no more than 25% of the total radiated power as surface waves.

The most important of the results published is by Pozar [1983]. He tested radiation efficiency against normalised substrate thickness. The radiation efficiency data were calculated using a moment method for a printed rectangular radiating element on a grounded dielectric slab. The moment method uses the rigorous dyadic Green's function for the grounded dielectric slab and so includes the exterior fields making calculations for surface wave excitation and mutual coupling possible. Pozar found that the surface wave excitation is generally not important for thin substrates, normally of the order of $h < 0.01\lambda_0$. An interesting observation was the similarity between the radiation efficiency for the dipole and the patch. It was also shown that the radiation efficiency does not depend on the patch width W or the feed location of the dipole or the patch.

An approach to the analysis of microstrip antennas using Green's function, which is applicable also to relatively thick substrates, was presented by Perlmutter *et al.* [1985]. This approach resembles the methods of Uzunoglu *et al.* [1979] and Van der Paw [1977] used for various other problems. A certain current distribution was assumed along the upper conductor that is typical of the geometry of the element. The current in the radiating element was obtained by using cavity or equivalent transmission line models. The surface wave excitation was then found from the assumed currents using the appropriate Green's function in the Fourier domain. It has been shown that increasing h causes a larger fraction of the power to be coupled into surface waves [Perlmutter *et al.*, 1985]. However, this fraction is approximately independent of the patch width W.

Mosig and Gardiol [1985] presented a dynamic analysis of microstrip structures. It was shown that the mixed-potential integral equation for stratified media provides a rigorous and powerful approach. Green's functions belonging to the kernel of the integral equation were expressed as Sommerfeld integrals, in which surface wave effects are automatically included.

Bhattacharyya and Garg [1986] proposed a general approach for the determination of power radiated via the space and surface waves from the aperture of an arbitrarily-shaped microstrip antenna. The magnetic current model was used and the analysis was carried out in the Fourier domain to determine the effect on the substrate. It was observed that for $h/\lambda_d < 0.02$, where λ_d is the wavelength in the substrate, the effect of surface waves can be ignored. The results obtained by Bhattacharyya and Garg [1986] confirmed the results obtained by Pozar [1983] but did not provide extra material.

In [Jackson and Alexopoulos, 1991], an approximate formula was derived for the radiation efficiency of a resonant rectangular microstrip patch. This formula came from approximations of a rigorous Sommerfeld solution and was not empirical. Jackson and Alexopoulos [1991] also showed that the radiation efficiency decreases much more rapidly with increasing substrate thickness when using a magnetic substrate. Additionally, for $h/\lambda_o \geq 0.05$, the radiation efficiency results calculated from the approximate formula do not correspond well with the exact results obtained from a rigorous Sommerfeld solution.

From these studies, the rigorous way of calculating the radiation efficiency of rectangular microstrip antennas involves complicated Green's function methods and integral transformation techniques.

This section describes the derivation of a new simple formula for the radiation efficiency of rectangular microstrip antennas using tabu search. This formula shows explicitly the dependence of radiation efficiency on the characteristic parameters of a patch antenna. Thus, the radiation efficiency of rectangular microstrip antennas can be accurately and easily calculated using this formula, without requiring complicated Green's function methods or integral transformation techniques. The results obtained from this formula agree well with the results available in the literature, even when $h(\varepsilon_r)^{1/2}/\lambda_0 = 0.31$.

3.2.1 Radiation Efficiency of Rectangular Microstrip Antennas

Consider a rectangular patch of width W and length L over a ground plane with a substrate of thickness h and a relative dielectric constant ε_r, as shown in Figure 3.2. The radiation efficiency due to surface waves is defined as follows:

$$\eta = \frac{P_{sp}}{P_{sp} + P_{su}} \qquad (3.10)$$

where P_{sp} is the power radiated in space waves and P_{su} is the power radiated in surface waves. $P_{sp} + P_{su}$ is then the total power delivered to the printed antenna element. Although P_{sp} is easily found, P_{su} has to be obtained by complicated Green function methods.

In this study, to determine the radiation efficiency of a rectangular microstrip antenna, the radiation efficiency results reported by Pozar [1983] and Perlmutter *et al.* [1985] will be concentrated upon because their results agree with those presented by other researchers. The results calculated by Pozar [1983] using a moment method for a substrate with relative permittivity ε_r equal to 12.8 are given in Figure 3.3. The results calculated according to Perlmutter *et al.* [1985] using the electric surface current model are presented in Figures 3.4-3.5 for $\varepsilon_r = 2.2$ and 9.8. From these plots, it can be seen that the radiation efficiency decreases as the substrate thickness increases. This is because the surface wave power increases while the space wave power is reduced. The curves also indicate that a lower value of ε_r results in a higher efficiency. Figures 3.3-3.5 show that the width W of the patch has almost no effect on the value of η. The difference between the η values for a patch (dotted curve) and a dipole (broken curve) is less than 0.02 for $\varepsilon_r = 2.2$ and almost zero for $\varepsilon_r = 12.8$. As only resonant antennas are of interest, the physical length L of the patch is not important. It is determined by:

$$L = \frac{c}{2 f_r \sqrt{\varepsilon_e}} - 2\Delta L \tag{3.11}$$

where c is the velocity of electromagnetic waves in free space, ε_e is the effective relative dielectric constant for the patch, f_r is the resonant frequency and ΔL is the edge extension. ε_e and ΔL depend on ε_r, h and W. Thus, the length L is determined by W, h, ε_r and f_r and only two parameters are needed to describe the radiation efficiency, namely ε_r and h/λ_0. The problem is that a simple formula for the radiation efficiency must be obtained. However, the results obtained from using the formula must correspond well with the results produced by more complicated methods, such as Green function methods or integral transformation techniques.

3.2.2 Application of Tabu Search to the Problem

First, a model for the formula is chosen and then the optimal coefficient values of the model are found using tabu search. To obtain a suitable model for the radiation efficiency formula, many experiments were carried out and the following model was chosen:

$$\eta = 1 + \alpha_1 F^{\alpha_2} G^{\alpha_3} \varepsilon_r^{\alpha_4} + \alpha_5 F^{\alpha_6} G^{\alpha_7} \varepsilon_r^{\alpha_8} \tag{3.12}$$

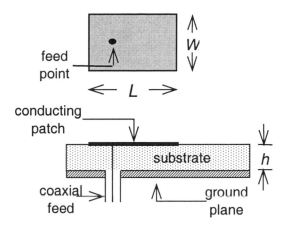

Figure 3.2 Geometry of rectangular microstrip antenna

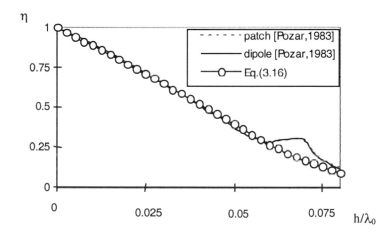

Fig. 3.3 Radiation efficiency for patch antenna and dipole on substrate with $\varepsilon_r = 12.8$

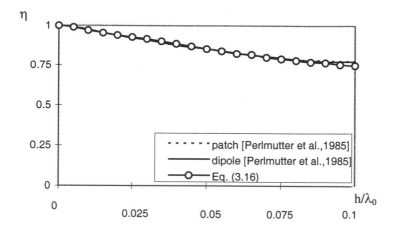

Fig. 3.4 Radiation efficiency for wide and narrow patch antenna on substrate with $\varepsilon_r = 2.2$

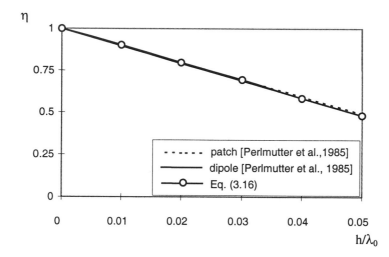

Fig. 3.5 Radiation wave efficiency for wide and narrow patch antenna on substrate with $\varepsilon_r = 9.8$

where $F = (\varepsilon_r - 1)$ and $G = h/\lambda_0$. It is clear from equation (3.12) that the model depends on ε_r and h/λ_0. The coefficients $\alpha_1 ... \alpha_i ... \alpha_8$ are determined by the tabu search algorithm. The term F in equation (3.12) ensures that $\eta = 1$ for an air dielectric.

A solution is represented as a string containing 8 real numbers (coefficient values) and has an associated set of neighbours. The initial solution used by the tabu search algorithm at the start consists of randomly produced coefficient values. Coefficients can have positive or negative values between -5 and 5. A neighbour of the present solution is produced by adding a randomly generated number between ±1 to a non-tabu coefficient of the present solution. Hence, for each iteration, the maximum number of neighbours to be produced is 8. The recency and frequency limits used in this work for tabu restrictions are:

recency limit = 0.75 × number of coefficients
frequency limit = 1.5 × average frequency   (3.13)

where average frequency is the average rate of change of all coefficient values.

To calculate the performance of a neighbour, first equation (3.12) is established using the coefficient values obtained from the neighbour x^*. Second, η is computed using this formula for three different values of ε_r and a predetermined interval of (h/λ_0). Next, the performance of the neighbour is calculated by the following formula:

$$p(x^*) = A - (1/N) \sum_{j=1}^{N} (\eta_t(j) - \eta(j))^2 \qquad (3.14)$$

where A is a positive constant selected to be large enough so that the p value is positive for all possible solutions, N is the total number of efficiency values employed for the optimisation process, and η_t and η represent, respectively, the radiation efficiency values obtained by the well-known Green function methods [Pozar, 1983; Perlmutter *et al.*, 1985] and by equation (3.12) established using the coefficient values produced from neighbour x^*.

Lastly, the performance values of all neighbours are compared and the neighbour which produces the maximum performance is selected as the next solution. This process is repeated until a given stopping criterion, such as the number of iterations being equal to a preset maximum, is satisfied.

3.2.3 Simulation Results and Discussion

In the optimisation process, the data sets obtained from the methods of Pozar [1983] and Perlmutter *et al.* [1985] for different dielectric permittivities and substrate thicknesses were used. In the test stage, η was computed by the proposed formula obtained using tabu search for unseen data sets in the optimisation stage. 107 data sets consisting of ε_r, h/λ_0 and η_t values used for the optimisation process were generated from the moment method [Pozar, 1983] for $\varepsilon_r = 12.8$ and from the electric surface current model [Perlmutter *et al.*, 1985] for $\varepsilon_r = 2.2$ and 4.0.

The tabu search algorithm was run for 400 iterations and the following optimum values for the unknown coefficients of the model given in equation (3.12) were found:

$$\alpha_1 = -3.66 \quad \alpha_2 = 1.83 \quad \alpha_3 = 1.06 \quad \alpha_4 = -1.32$$
$$\alpha_5 = -2.48 \quad \alpha_6 = 2.48 \quad \alpha_7 = 0.5 \quad \alpha_8 = -3.12 \tag{3.15}$$

The following radiation efficiency formula is then obtained by substituting these coefficient values into equation (3.12):

$$\eta = 1 - 3.66 F^{1.83} G^{1.06} \varepsilon_r^{-1.32} - 2.48 F^{2.48} G^{0.5} \varepsilon_r^{-3.12} \tag{3.16}$$

Figures 3.3-3.5 show the optimisation results. The solid curves in these figures represent the results obtained by using equation (3.16). It can be seen that the results of the proposed formula were close to the results of Green function methods.

To test the proposed radiation efficiency formula, in equation (3.16), the results of the electric surface current model for $\varepsilon_r = 9.8$ and for different substrate thicknesses which were not used in the optimisation process are compared with the results of the equation. This comparison is shown in Figure 3.5. It is apparent from the figure that the radiation efficiency for wide and narrow patch antennas on substrates with different thicknesses is computed to a high accuracy.

Both optimisation and test results illustrate that the performance of the formula is quite robust and precise. When the results of the proposed formula are compared with those of Pozar [1983] and Perlmutter *et al.* [1985], the error is within 0.023, which is tolerable for most design applications.

The data sets for $\varepsilon_r = 2.2$, 9.8 and 12.8 were also used in the optimisation process and the data set for $\varepsilon_r = 4.0$ in the test process. In this case, the same model given in equation (3.12) was chosen and the different coefficient values were found. It was observed again that the results obtained agree well with the results of Green function methods.

It should be emphasised again that more accurate results could be obtained using higher-order models but at the expense of formula simplicity. It seems practical to use a simple formula that lends insight into the dependence of the radiation efficiency upon the various parameters, such as thickness h and relative dielectric constant ε_r. Even though the formula is simple, it provides accurate results in many cases.

As the difference between radiation efficiency for the dipoles and patches is always less than 0.02, equation (3.16) can also be used for dipoles.

Since the formula presented in this work has a high accuracy, in the range $1 \leq \varepsilon_r \leq 12.8$ and $0 < h/\lambda_d \leq 0.31$, and requires no complicated mathematical functions, it can be very useful for the development of fast CAD algorithms. Using this formula, one can calculate the radiation efficiency of rectangular patch antennas by hand, without possessing any background knowledge of microstrip antennas. The formula given by equation (3.16) can also be used for many other engineering applications.

3.3 Training Recurrent Neural Networks for System Identification

Tabu search is good at local convergence, due to its iterative nature, but can have problems reaching the global optimum solution in a reasonable computation time, particularly when the initial solution is far from the global optimum solution [Karaboga and Kaplan, 1995; Karaboga and Kalinli, 1996a].

The convergence speed of tabu search can be improved by introducing parallelism [Malek *et al.*, 1989]. This helps the tabu search to find promising regions of the search space more quickly. In this section, a parallel model for tabu search is described together with its application to the training of a recurrent neural network for dynamic system identification [Karaboga and Kalinli; 1996a, 1996b, 1997].

3.3.1 Parallel Tabu Search

The flowchart of the parallel tabu search (PTS) algorithm is depicted in Figure 3.6. Sub-blocks 1 to n are each identical to the flowchart shown in Figure 1.5.

These blocks represent *n* tabu searches (TSs) executing independently. The initial solutions for the searches during the first cycle are created using a random number generator with different seeds.

After a given number of iterations, *maxit*, the execution of the tabu searches is stopped. *maxit* is normally chosen to be sufficiently large to allow the tabu search to complete local searching. An operation is applied to the solutions found by *n* parallel TSs during the first cycle to create the initial solutions for TSs for the second cycle. Various methods could be used for this purpose. For example, the best solution found by the TSs during the first cycle could be selected as one of the initial solutions. The others could be produced by applying a suitable crossover operator to the solutions obtained in the first cycle. Alternatively, all the initial solutions could be created using the crossover operator. Another possibility is that that the best solution is selected directly, others are created by applying the crossover operator to the solutions obtained in the first cycle and the rest by using a random number generator. The crossover operation is explained in the following section.

After the final solutions are obtained from the *n* tabu search algorithms, the so-called *Elite* unit chooses the best of the solutions to be the optimal solution.

3.3.2 Crossover Operator

In this work, the crossover operator of [Michalewicz, 1992] is employed. Two solutions (W and V) are taken. From these, a new solution is generated which is the average of both solutions. This is illustrated below:

Present solution 1 | W_1 | W_2 | | W_i | | W_n |

Present solution 2 | V_1 | V_2 | | V_i | | V_n |

New solution | $0.5(W_1+V_1)$ | $0.5(W_2+V_2)$ | | $0.5(W_i+V_i)$ | | $0.5(W_n+V_n)$ |

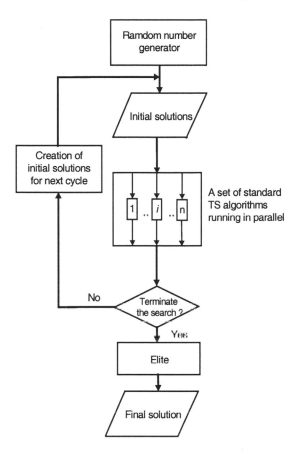

Fig. 3.6 Structure of the parallel tabu search algorithm

3.3.3 Training the Elman Network

The Elman network employed is shown in Figure 2.23 [Elman, 1990]. The structure of the network is as used in [Karaboga and Kalinli, 1997; Pham and Liu, 1997] to facilitate comparison of results. Since the plants to be identified are single-input single-output, the number of input and output neurons is one. There are six neurons in the hidden layer. The total number of connections is fifty-four, of which six are feedback connections.

A solution is represented as a string of trainable weights. When all connections are trainable, feedback connections have weight values in the range 0.0 to 1.0 while feedforward ones can have positive or negative weights between 1.0 and -1.0. When the feedback connection weights are constant, their values are taken as 1.0 ($\alpha_i = 1$). A neighbour solution is produced from the present solution by adding a random number between -1.0 and 1.0 to a trainable weight that is not tabu. Note that from the point of view of the TS algorithm, there is no difference between the feedback and feedforward connections and training of one type of connections is carried out identically to the other.

In the training stage, first, a sequence of the input signals $u(k)$, $(k = 0,1....)$ is fed to both the plant and the recurrent network designed with the weights obtained from a neighbour. Second, the RMS error value between the plant and recurrent network outputs is computed. Next, according to the RMS error values computed for the neighbours, the next solution is selected. The neighbour solution that produces the minimum RMS error value is selected as the next solution.

The recency and frequency limits are taken as:

recency limit = number of trainable weights×0.5
frequency limit = average frequency×1.5

3.3.4 Simulation Results and Discussion

Simulations were conducted to study the ability of the RNN trained by the standard TS algorithm and the proposed TS algorithm to model linear and non-linear plants. The linear plant was a third-order system with the following discrete-time equation:

$$y(k) = A_1 y(k-1) + A_2 y(k-2) + A_3 y(k-3) + B_1 u(k-1) + B_2 u(k-2) + B_3 u(k-3) \qquad (3.17)$$

where $A_1= 2.627771$, $A_2= -2.333261$, $A_3= 0.697676$, $B_1= 0.017203$, $B_2 = -0.030862$ and $B_3 = 0.014086$.

An Elman network using all linear neurons was tested. The training input signal $u(k)$, $k = 0,1,...,99$, was random and varied between -1.0 and 1.0. First, the results were obtained by assuming that only the feedforward connection weights were trainable. Second, results were produced for the Elman network with all connection weights variable. For each case, experiments were repeated six times for different initial solutions. The results obtained by using the standard BP and TS algorithms and the proposed TS algorithm are given in Figure 3.7 and the responses are presented in Figures 3.8 and 3.9, respectively. PTS in Figure 3.7 refers to the proposed TS algorithm and TS the standard algorithm.

The non-linear model adopted for the simulations was that of a simple pendulum swinging through small angles [Liu, 1993]. The discrete-time description of the plant is:

$$y(k) = A_1 y(k-1) + A_2 y(k-2) + A_3 y^3(k-2) + B_1 u(k-2) \qquad (3.18)$$

where A_1=1.04, A_2= -0.824, A_3= 0.130667 and B_1 = -0.16.

The Elman network with non-linear neurons in the hidden layer was employed. The hyperbolic tangent function was adopted as the activation function of the non-linear neurons.

The neural networks were trained using the same sequence of random input signals as mentioned above. As in the case of the linear plant, the results were obtained for six different runs with different initial solutions. The RMS error values obtained by the standard BP and TS algorithms and the proposed TS algorithm are presented in Figure 3.10. The responses of the non-linear plant and the recurrent network with the weights obtained by the standard and proposed TS algorithms are shown in Figures 3.11 and 3.12, respectively.

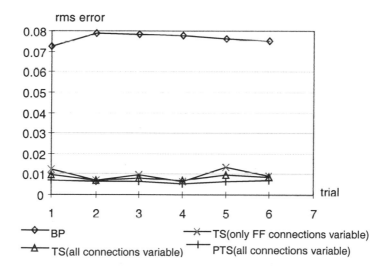

Fig. 3.7 RMS error values obtained for the linear plant for six runs with different initial solutions

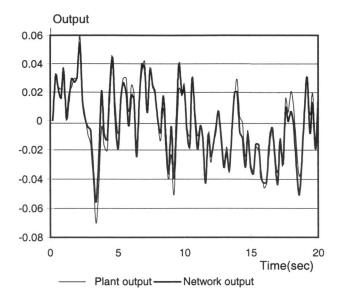

Fig. 3.8 Responses of the plant and the network trained by the standard tabu search (third-order linear plant) (RMS error = 0.0093218)

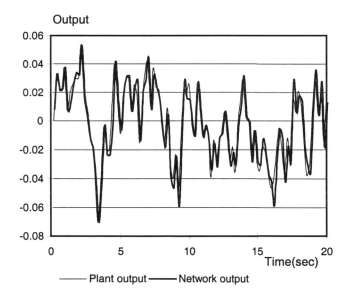

Fig. 3.9 Responses of the plant and the network trained by the proposed tabu search (third-order linear plant) (RMS error = 0.0054352)

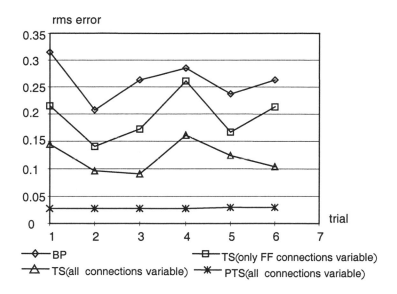

Fig. 3.10 RMS error values obtained for non-linear plant for six different runs with different initial solutions

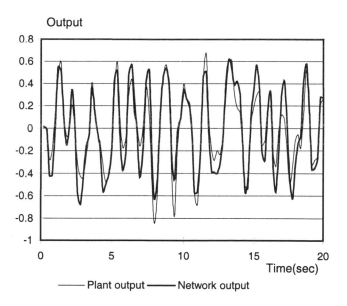

Fig. 3.11 Responses of the non-linear plant and the network trained by the standard tabu search (RMS error = 0.141587)

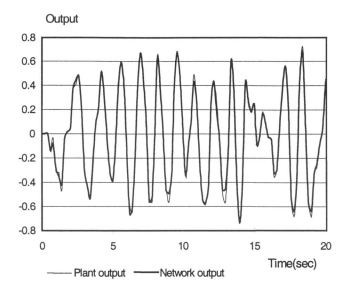

Fig. 3.12 Responses of the non-linear plant and the network trained by the proposed tabu search (RMS error = 0.02714256)

The proposed TS algorithm was executed for seven cycles. The number of TSs running in parallel was four (n=4). The parameters of all TSs were the same. The total number of evaluations was 24000 since each TS was run for 6000 evaluations.

These results show that the Elman network trained by TSs could identify the third-order linear plant successfully. Note that an original Elman network with an identical structure to that adopted in this work and trained using the standard backpropagation algorithm failed to identify even a second-order linear plant [Liu, 1993]. Moreover, when the Elman network had been trained by a standard genetic algorithm [Goldberg, 1989], the third-order plant could not be identified although the second-order plant was [Pham and Karaboga, 1999]. It is clear from Figures 3.7-3.12 that the proposed tabu search algorithm trained the networks better than the standard BP and TS algorithms for both network structures when all connection weights were variable and when only feedforward connection weights were trainable. The strong performance of the proposed TS is more evident in the case of identifying the non-linear plant while the performance of both TSs is similar for identifying the linear plant.

3.4 Designing Digital Finite-Impulse-Response Filters

Generally, the implementation of a digital filter is carried out with hardware, which gives rise to an important practical design constraint. This is that each theoretical infinite precision coefficient used to describe the filter must be represented by a finite number of bits. Simple rounding or truncating of each coefficient to a finite wordlength (FWL) coefficient produces a filter that may have a sub-optimal set of coefficients. Several computer techniques have been developed for determining an optimal set of FWL coefficients from the infinite set of precision coefficients [Diethorn and Munson, 1986; Karaboga *et al.*, 1995; Kodek, 1980; Xu and Daley, 1992]. These techniques employ a search method using probabilistic rules to produce FWL coefficients. As the number of FWL coefficients used to describe the filter increases, the computation time increases significantly for minimising the difference between the desired frequency response and that of the digital filter described by a set of FWL coefficients. For example, with the use of a general purpose integer-programming algorithm, the computation time increases exponentially with the number of FWL coefficients. In the case of methods based on genetic algorithms, promising regions in the search space can be found very quickly because of the parallel nature of the algorithm. However, since the same solutions might be evaluated several times, due to the probabilistic transition rules employed, reaching an optimal solution may take a long time [Karaboga *et al.*, 1995].

In this section, a novel technique based on parallel tabu search is described to design an optimal FWL finite impulse response (FIR) filter.

3.4.1 FIR Filter Design Problem

The transfer function of an Nth order FIR filter is:

$$H(f) = \sum_{i=0}^{N-1} a_i \cos(2\pi f) \tag{3.19}$$

The FWL coefficients a_i are assumed to be represented by:

$$a_i = \sum_{j=1}^{M} b_{ij} 2^j - b_{i0} \tag{3.20}$$

where $b_{ij} \in \{0,1\}$ and M is an integer. The FIR filter design problem can be viewed as the optimisation of the error function $E(f)$ for a given filter specification in the frequency range $f \in [0,...0.5]$. Here, $E(f)$ is assumed to be defined by:

$$-\delta_{f_i} \le E(f_i) = \left[W(f_i) \left(\left| H(f_i) \right| - \left| H_d(f_i) \right| \right) \right] \le \delta_{f_i} \qquad (3.21)$$

where $|H_d(f)|$ and $|H(f)|$ are the desired and the actual magnitude responses, respectively, and $W(f)$ is a specified error-weighting function.

3.4.2 Solution by Tabu Search

The PTS algorithm, described in the previous section, is employed. The crossover operator used is the single-point crossover described in Chapter 1. The standard TSs utilised in the first stage are initialised to a solution obtained by rounding the set of infinite length coefficients found using the Parks-McClellan algorithm to the nearest finite wordlength number. After a specified number of iterations, the solutions found by the standard TSs in the first cycle are collected in a pool and the crossover operation is applied to them to produce a set of new solutions. The new solutions obtained after the crossover operation are taken as the initial solutions for the standard TSs existing at the next stage. This process is repeated for a predetermined number of cycles. Here, the number of cycles is seven, the iteration number for each standard TS is 400 and the number of PTSs is 5.

For this problem, the TS algorithm searches with coefficient moves restricted to integers between ±2. Each solution is represented as a string of integer numbers and a neighbour solution is produced by adding an integer between ±2 to the associated rounded coefficient. Hence, for each coefficient that is not tabu, four neighbours are created and evaluated. After this process is carried out for all non-tabu coefficients, the next solution is selected from the neighbours. The performance of a neighbour is calculated using the following equation:

$$P_i = Max \left\{ W(f) \left[\left| H_i(f) \right| - \left| H_d(f) \right| \right] \right\}^{-2} \qquad (3.22)$$

where P_i is the performance value of the ith neighbour and $|H_i|$ is the actual response produced by the filter designed using the coefficients obtained from the ith neighbour. If there are neighbours for which the performance values are higher than that of the present solution, the best neighbour is selected as the next solution. If none of the neighbours is better than the present solution, the next solution is determined according to the evaluation values obtained by the following formula:

$$e_i = a\,P_i + b\,r_i - c\,f_i \qquad (3.23)$$

where P_i, r_i and f_i are the performance, recency and frequency values of the ith neighbour, respectively. In equation (3.23), a, b and c are the performance, recency and frequency factors and are assigned the values 0.1, 2 and 1, respectively.

After the next solution is determined, the past moves memory is modified. For other iterations, the same procedure is repeated. If all neighbours are tabu, the tabu move that loses its tabu status by the least increase in the value of current iteration is freed from the tabu list.

3.4.3 Simulation Results

The desired frequency response of the filter is taken as:

$$H_d(f) = \begin{cases} 1 & f \in [0,0.2] \\ 0 & f \in [0.25,0.5] \end{cases} \tag{3.24}$$

with the weighting function defined to be unity. The length of the filter is 40. The results of the tabu search are shown in Table 3.5. The performance characteristics of the filter designed by these techniques are also summarised in Table 3.6.

In this application, the design with the parallel tabu search algorithm produces the least error in the stopband and passband. This design exceeds the performance of those obtained with the integer-programming and round-off methods previously used.

The magnitude frequency response characteristics for each of these filter designs are shown in Figure 3.13. All filter characteristics are very similar in the passband. However, in the stopband, the filter designed by the proposed algorithm produces the best performance and the greatest attenuation. It is seen that the proposed procedure performs better than simple rounding by almost 5dB and than integer programming by nearly 3dB in the stopband.

Table 3.5 Filter coefficients obtained by using three different design methods

Coefficients	Integer programming	Round-off	PTS
h(0)= h(39)	2	1	2
h(1)= h(38)	2	4	2
h(2)= h(37)	-1	-2	-1
h(3)= h(36)	-4	-4	-4
h(4)= h(35)	0	0	0
h(5)= h(34)	6	6	5
h(6)= h(33)	1	2	1
h(7)= h(32)	-8	-7	-7
h(8)= h(31)	-5	-5	-5
h(9)= h(30)	8	8	8
h(10)= h(29)	10	10	10
h(11)= h(28)	-9	-8	-8
h(12)= h(27)	-17	-17	-16
h(13)= h(26)	5	5	5
h(14)= h(25)	27	27	27
h(15)= h(24)	2	3	3
h(16)= h(23)	-43	-44	-43
h(17)= h(22)	-25	-24	-25
h(18)= h(21)	92	92	93
h(19)= h(20)	211	212	212

Table 3.6 Performance characteristics of the filter designed by three methods

Methods	Passband deviation	Stopband deviation (dB)
Round-off	0.020087	0.01747(-35.15dB)
Integer programming	0.016144	0.01380(-37.20dB)
PTS	0.014304	0.01011(-39.90dB)

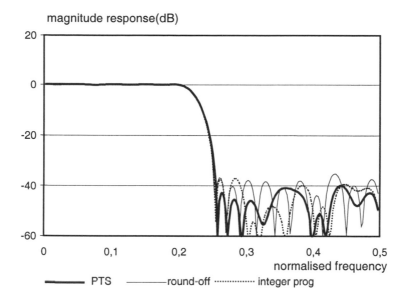

Fig. 3.13 Frequency response characteristics of the filters designed by three different methods

3.5 Tuning PID Controller Parameters

The best-known controllers used for industrial control processes are PID controllers since their structure is simple and the performance is robust for a wide range of operating conditions. In the design of such controllers, three control parameters have to be specified: proportional, integral and derivative gains. The control parameters of a standard PID controller are fixed during control after they have been chosen. Several methods have been developed to produce appropriate values for the parameters [Ziegler and Nichols, 1942; Kitamori, 1979; Hang *et al.*, 1991]. Depending on the nature of the process to be controlled, non-linearity and order etc., the derivation of appropriate values for the parameters may be very difficult. In an adaptive PID controller design, another problem arises since the time available to optimise the parameters is limited.

Let q be a vector containing the three PID parameters. The space of possible vectors q is potentially very large. The design of the PID controller can be regarded as the selection of an optimal vector q' that will minimise a cost function $J(T_c)$ subject to the constraint that q' belongs to the constrained design space Q. T_c

is the total time taken to find the parameter values. In general, the constrained design space Q is a complex multidimensional space. To find the optimal vector q', a robust search algorithm is required to explore the space efficiently.

This section describes a new method based on tabu search for tuning PID control parameters.

3.5.1 Application of Tabu Search to the Problem

The discrete-time equivalent expression for the output of the PID controller is:

$$u(k) = K_p e(k) + K_i T_s \sum_{i=1}^{n} e(i) + (K_d/T_s) \Delta e(k) \qquad (3.25)$$

where K_p, K_i and K_d are the proportional, integral and derivative gains, respectively. K_i and K_d can be stated in terms of K_p in the following form:

$$K_i = K_p/T_i , \ K_d = K_p T_d \qquad (3.26)$$

Here, T_i and T_d are known as the integral and derivative time constants, respectively.

In equation (3.25), $u(k)$ and $e(k)$ are, respectively, the control signal and error between the reference and the output. T_s is the sampling period for the controller.

It is assumed that the control parameters K_p, K_i and K_d are in predetermined ranges $[K_{pmin}, K_{pmax}]$, $[K_{imin}, K_{imax}]$ and $[K_{dmin}, K_{dmax}]$, respectively. Then, the tabu search algorithm seeks optimal values of the parameters within these ranges. Appropriate ranges for the parameters are defined by the following relations [Zhao *et al.*, 1993]:

$$K_{pmin} = 0.32K_u, \qquad K_{pmax} = 0.6 \ K_u$$

$$K_{dmin} = 0.08K_u T_u \qquad K_{dmax} = 0.15 \ K_u \ T_u \qquad (3.27)$$

$$K_{imin} = K^2_{pmin}/ \ 3K_{dmin} \qquad K_{imax} = K^2_{pmax}/ \ 3K_{dmin}$$

where K_u and T_u are, respectively, the gain and the period of oscillation at the stability limit under P-control.

The cost function to be minimised is:

$$J(T_c) = \sum_{k=1}^{T_c} (\mid y_d(k) - y_p(k) \mid kT) \tag{3.28}$$

where y_d and y_p are the desired and actual outputs of the plant, respectively, T_c is the total time over which the cost is computed and T is the sampling period.

The performance of the PID controller designed by the tabu search algorithm is calculated using its cost value. For this purpose the following function is used:

$$perf(i) = A\text{-}J(i) \tag{3.29}$$

where A is a positive constant and $J(i)$ is the cost value computed when the parameter values obtained from the ith move or neighbour of the present solution are used. Here, A is 100.

For the forbidding strategy of the tabu search algorithm, two conditions are used:

$recency(i) <- r_1 N$ (recency condition)
$frequency(i) >= f_1 avfreq$ (frequency condition)

Here, r_1 and f_1 are recency and frequency factors for the forbidding strategy and are equal to 0.5 and 1.5. N is the number of elements in the binary sequence and *avfreq* is the average rate of change of bits.

If none of the neighbours produces better performance than the present solution, the next solution is selected according to the evaluation values calculated by:

$$eval(i) = a \times improvement(i) + r_2 \times recency(i) - f_2 \times frequency(i) \tag{3.30}$$

where a, r_2 and f_2 are the improvement, recency and frequency factors used for the evaluation function, respectively. These factor values are kept as 3, 2 and 1.5, respectively. *improvement(i)* is the difference between the performance values of the present solution s and its neighbour (i).

$$improvement(i) = perf(i) - perf(s) \tag{3.31}$$

3.5.2 Simulation Results

Here, eight bits are employed to represent each parameter. N is equal to 24 and the length of a binary sequence is 24 bits. Each parameter is first generated as an integer and then scaled into the predefined range.

The proposed method has been tested on the following time-delayed second-order process for comparison [Zhao *et al.*, 1993].

$$G(s) = e^{-0.5s} / (s+1)^2 \tag{3.32}$$

Table 3.7 shows the simulation results obtained. In the table, y_{os} represents the percentage of maximum overshoot, T_s stands for the 5% settling time and IAE, ISE are the integral of the absolute error and the integral of the squared error, respectively. The time response obtained from the process under the control of the PID controller designed using tabu search is plotted in Figure 3.14.

The results obtained by using the Ziegler-Nichols PID and Kitamori's PID controllers are also presented for comparison. The parameters of the Ziegler-Nichols PID controller were determined as $K_p = 0.6K_u$, $T_i = 0.5T_u$ and $T_d = 0.125T_u$. The controller designed using the proposed method yields better control performance than Ziegler-Nichols' and Kitamori's PID controllers. This can be seen from the performance index values given in Table 3.7.

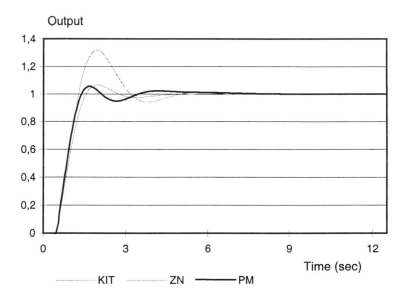

Fig. 3.14 Time responses obtained from the process under the control of controllers designed by the Ziegler-Nichols method (ZN), the Kitamori mthod (KIT) and the proposed method (PM)

Table 3.7 Results obtained by using Ziegler-Nichols, Kitamori and proposed PID controllers

Ziegler-Nichols' PID controller	Kitamor's PID controller	Proposed PID controller
K_p=2.808	K_p=2.212	K_p=2.194
K_d= 1.151	K_d= 1.148	K_d= 1.465
K_i= 1.7122	K_i= 1.0858	K_i= 1.209
IAE= 1.375	IAE= 1.016	IAE= 0.995
ISE= 0.842	ISE= 0.787	ISE= 0.741
Y_{os} =32%	Y_{os} =6.7%	Y_{os} =5.5%
T_s =4.13	T_s =2.304	T_s = 1.8

3.6 Summary

This chapter has described five different applications of tabu search. In the first application, a new and very simple curve-fitting expression for the resonant frequency of triangular microstrip antennas in terms of the effective side length is obtained using a modified tabu search algorithm. The theoretical resonant frequency results calculated using this new side-length expression correspond well with the experimental results available in the literature.

The second application involves obtaining a simple formula for the radiation efficiency of rectangular microstrip antennas. The good correlation between the results produced from the proposed formula and the more complex Green function methods supports the validity of the formula.

The third application employs a new parallel tabu search model adopting the crossover operator of genetic algorithms. The application concerns the training of Elman recurrent neural networks to identify linear and non-linear plants. The simulation results obtained show that the proposed TS model can be used for training Elman networks to identify linear and non-linear plants efficiently.
The fourth application is the design of digital FIR filters using the parallel tabu search algorithm adopted in the previous application. This section gives an illustrative design example that shows that the proposed procedure can provide a good solution.

The fifth application is tuning the control parameters of a PID controller for a given process. Simulation results for a time-delayed second-order process demonstrate the good performance of the proposed method compared with that of two well-known design methods.

References

Bahl, I.J. and Bhartia, P. (1980) *Microstrip Antennas,* Artech House, Dedham, MA.

Bhattacharyya, A.K. and Garg, R. (1986) Effect of substrate on the efficiency of an arbitrarily shaped microstrip patch antenna, *IEEE Trans. Antennas Propagation*, Vol.AP-34, pp.1181-1188.

Carver, K.R. and Mink, J.W. (1981) Microstrip antenna technology, *IEEE Trans. Antennas Propagation*, Vol.AP-29, pp.2-24.

Chen, W., Lee, K.F. and Dahele, J.S. (1992) Theoretical and experimental studies of the resonant frequencies of the equilateral triangular microstrip antenna, *IEEE Trans. Antennas Propagation*, Vol.AP-40, No.10, pp.1253-1256.

Dahele, J.S. and Lee, K.F. (1987) On the resonant frequencies of the triangular patch antenna, *IEEE Trans. Antennas Propagation*, Vol.AP-35, No.1, pp.100-101.

Diethorn, E.J. and Munson, D.C. (1986) Finite wordlength FIR digital filter design using simulated annealing, *Proc. Int. Symp. Circuits and Systems*, pp.217-220.

Elman, J.L. (1990) Finding structure in time, *Cognitive Science*, Vol.14, pp.179-211.

Gang, X. (1989) On the resonant frequencies of microstrip antennas, *IEEE Trans. Antennas Propagation*, Vol.AP-37, No.2, pp.245-247.

Garg, R. and Long, S.A. (1988) An improved formula for the resonant frequency of the triangular microstrip patch antenna, *IEEE Trans. Antennas Propagation*, Vol. AP-36, p.570.

Glover, F. (1989) Tabu Search Part I, *ORSA J. on Computing*, Vol.1, No.3, pp.190-206.

Glover, F. (1990) Tabu Search Part II, *ORSA J. on Computing*, Vol.2, No.1, pp.4-32.

Goldberg, D.E. (1989) *Genetic Algorithms in Search, Optimisation and Machine Learning,* Addison-Wesley, Reading, Mass.

Güney, K. (1993a) A new edge extension expression for the resonant frequency of electrically thick rectangular microstrip antennas, *Int. J. of Electronics*, Vol.75, No.4, pp.767-770.

Güney, K. (1993b) Resonant frequency of a triangular microstrip antenna, *Microwave and Optical Technology Letters*, Vol.6, No.9, pp.555-557.

Güney, K. (1993c) Space-wave efficiency of rectangular microstrip antennas, *Int. J. of Electronics*, Vol.74, pp.765-769.

Güney, K. (1994a) Comments on the resonant frequencies of microstrip antennas, *IEEE Trans. Antennas Propagation*, Vol.AP-42, No.9, pp.1363-1365.

Güney, K. (1994b) Resonant frequency of a tunable rectangular microstrip patch antenna, *Microwave and Optical Technology Letters*, Vol.7, No.12, pp.581-585.

Güney, K. (1994c) Resonant frequency of electrically-thick circular microstrip antennas, *Int. J. of Electronics*, Vol.77, No.3, pp.377-386.

Güney, K. (1995) Space-wave efficiency of electrically thick circular microstrip antennas, *Int. J. of Electronics*, Vol.78, pp.571-579.

Hang, C.C., Aström, K.J. and Ho, W.K. (1991) Refinements of the Ziegler-Nichols tuning formula, *Proc. IEE Pt.D.,* Vol.138, pp.111-118.

Helszajn, J. and James, D.S. (1978) Planar triangular resonators with magnetic walls, *IEEE Trans. Microwave Theory Tech.*, Vol.MTT-26, No.2, pp.95-100.

Jackson, D.R. and Alexopoulos, N.G. (1991) Simple approximate formulas for input resistance, bandwidth, and efficiency of a resonant rectangular patch, *IEEE Trans. Antennas Propagation*, Vol.AP-39, pp.407-410.

James, J.R. and Hall, P.S. (1989) *Handbook of Microstrip Antennas*, IEE Electromagnetic Wave Series No.28, Vols.1 and 2, Peter Peregrinus, London.

James, J.R., Hall, P.S. and Wood, C. (1981) *Microstrip Antennas - Theory and Design*, IEE Electromagnetic Wave Series No.12, Peter Peregrinus, London.

James, J.R. and Henderson, A. (1979) High-frequency behaviour of microstrip open-circuit terminations, *IEE J. Microwaves, Optics, and Acoustics,* Vol.3, pp.205-218.

Karaboga, D., Guney, K., Kaplan, A. and Akdagli, A. (1997a) A new effective side length expression obtained using a modified tabu search algorithm for the resonant frequency of a triangular microstrip antenna, *Int. J. of Microwave and Millimeter-Wave Computer-Aided Engineering*.

Karaboga, N., Horrocks, D.H., Karaboga, D. and Alci, M. (1995) Design of FIR filters using genetic algorithms, *European Conf. on Circuits Theory and Design*, Turkey, Vol.2, pp.553-556.

Karaboga, D., Horrocks, D.H., Karaboga, N. and Kalinli, A. (1997b) Designing digital FIR filters using tabu search algorithm, *IEEE Int. Symp. on Circuits and Systems*, Hong Kong.

Karaboga, D. and Kalinli, A. (1996a) A new model for tabu search algorithm, *1st Turkish Symposium on Intelligent Manufacturing Systems*, Turkey, pp.168-175.

Karaboga, D. and Kalinli, A. (1996b) Training recurrent neural networks using tabu search algorithm, *5th Turkish Symp. on Artificial Intelligence and Neural Networks*, Turkey, pp.293-298.

Karaboga, D. and Kalinli, A. (1996c) Tuning PID controllers using tabu search algorithm, *IEEE Int. Conf. on Systems, Man and Cybernetics*, Beijing, China, Vol.1, pp.134-136.

Karaboga, D. and Kalinli, A. (1997) Training recurrent neural networks for dynamic system identification using parallel tabu search algorithm, *12th IEEE Int. Symp. on Intelligent Control*, Istanbul, Turkey.

Karaboga, D. and Kaplan, A. (1995) Optimising multivariable functions using tabu search algorithms, *10th Int. Symp. on Computer and Information Sciences*, Vol.II, Turkey, pp.793-799.

Kitamori, T. (1979) A method for control system design based upon partial knowledge about controlled process, *Trans. Society of Instrument and Control Engineers of Japan*, Vol.15, pp.549-555.

Kodek, D.M. (1980) Design of optimal finite wordlength FIR filter using integer programming techniques, *IEEE Trans. on Acoustics, Speech and Signal Processing*, Vol.ASSP-28, pp.304-308.

Kumprasert N. and Kiranon, W. (1994) Simple and accurate formula for the resonant frequency of the equilateral triangular microstrip patch antenna, *IEEE Trans. Antennas Propagation*, Vol.AP-42, No.8, pp.1178-1179.

Liu, X. (1993) *Modelling and Prediction Using Neural Networks*, PhD Thesis, School of Engineering, Cardiff University, UK.

Lo, Y.T., Solomon, D. and Richards, W.F. (1979) Theory and experiment on microstrip antennas, *IEEE Trans. Antennas Propagation*, Vol.AP-27, pp.137-145.

Lo, Y.T., Wright, S.M. and Davidovitz, M. (1989) Microstrip antennas, *Handbook of Microwave and Optical Components,* in K. Chang (ed.), Vol.1, Wiley, New York, pp.764-889.

Malek, M., Guruswamy, M., Pandya, P. and Owens, H. (1989) Serial and parallel simulated annealing and tabu search algorithms for the travelling salesman problem, *Annals of Operations Research*, Vol.1, pp.59-84.

Michalewicz, Z. (1992) *Genetic Algorithms+Data Structures = Evolution Programs*, Springer-Verlag, USA.

Mosig, J.R. and Gardiol, F.E. (1983) Dielectric losses, ohmic losses and surface wave effects in microstrip antennas, *Int. Symp. of International Union of Radio Science*, Santiago de Compostela, pp.425-428.

Mosig, J.R. and Gardiol, F.E. (1985) General integral equation formulation for microstrip antennas and scatterers, *IEE Proc. Pt.H*, Vol.132, pp.424-432.

Nauwelaers, B. and Van De Capelle, A. (1989) Surface wave losses of rectangular microstrip antennas, *Electronics Letters*, Vol.25, pp.696-697.

Perlmutter, P., Shtrikman, S. and Treves, D. (1985) Electric surface current model for the analysis of microstrip antennas with application to rectangular elements, *IEEE Trans. Antennas Propagation*, Vol.AP-33, pp.301-311.

Pham, D.T. and Karaboga, D. (1999) Training Elman and Jordan networks for system identification using genetic algorithms, *J. of Artificial Intelligence in Engineering*, Vol.13, No.2, pp.107–117.

Pham, D.T. and Liu, X. (1997) *Neural Networks for Identification, Prediction and Control* (3rd printing), Springer-Verlag, London.

Pozar, D.M. (1983) Considerations for millimeter wave printed antennas, *IEEE Trans. Antennas Propagation*, Vol.AP-31, pp.740-747.

Rana, I.E. and Alexopoulos, N.G. (1981) Current distribution and input impedance of printed dipoles, *IEEE Trans. Antennas Propagation*, Vol.AP-29, pp.99-105.

Roudot, B., Terret, C., Daniel, J.P., Pribetich, P. and Kennis, P. (1985) Fundamental surface-wave effects on microstrip antenna radiation, *Electronics Letters*, Vol.21, pp.1112-1114.

Singh, A. De R. and Yadava, R.S. (1991) Comments on an improved formula for the resonant frequency of the triangular microstrip patch antenna, *IEEE Trans. Antennas Propagation*, Vol.AP-39, No.9, pp.1443-1445.

Uzunoglu, N.K., Alexopoulos, N.G. and Fikioris, J.G.(1979) Radiation properties of microstrip dipoles, *IEEE Trans. Antennas Propagation*, Vol.AP-27, pp.853-858.

Van der Paw, L.J. (1977) The radiation of electromagnetic power by microstrip configurations, *IEEE Trans. Microwave Theory Tech.*, Vol.MTT-25, pp.719-725.

Wood, C. (1981) Analysis of microstrip circular patch antennas, *IEE Proc. Pt. H*, Vol.128, pp.69-76.

Xu, D.J. and Daley, M.L. (1992) Design of finite word length FIR digital filter using a parallel genetic algorithm, *IEEE Southeast Conf.*, Birmingham, USA, Vol.2, pp.834-837.

Zhao, Z.Y., Masayoshi, T. and Isaka, S. (1993) Fuzzy gain scheduling of PID controllers, *IEEE Trans. on Systems, Man and Cybernetics*, Vol.23, No.5, pp.1392-1398.

Ziegler, J.G. and Nichols, N.B. (1942) Optimum settings for automatic controllers, *Trans. ASME*, Vol.64, pp.759-768.

Zürcher, J.F. and Gardiol, F.E. (1995) *Broadband Patch Antennas*. Artech House, Dedham, MA.

Chapter 4

Simulated Annealing

This chapter describes the application of the simulated annealing algorithm to three different engineering problems. The first application is the optimal alignment of laser chip and optical fibre to minimise power loss [Barrere, 1997]. The second and third applications are inspection stations allocation and sequencing and economic lot-size production [Hassan, 1997].

4.1 Optimal Alignment of Laser Chip and Optical Fibre

This section describes how to use the simulated annealing algorithm for the optimal alignment of laser and optical fibre to minimise the power loss arising in laser - optical fibre connections. After the technical background information is given and the experimental setup explained, modifications made to the simulated annealing algorithm are described and the experimental results presented. For comparison, these results are set against those obtained using a fuzzy logic control system applied to the same problem.

4.1.1 Background

It is difficult to transmit information over long distances with small energy losses. With the normal means of transmission, such as coaxial cables and copper wires, there is a decrease in the power proportional to the length of the cable. To increase the power reached at the end of the transmission line, energy is introduced through repeaters placed along the link. In the case of optical fibre transmission, these repeaters can be placed further apart than for other means of transmission. Fibre optics can only transport modulated light and therefore the transmitter component needs to convert electrical signals into light and the receiver component has to accomplish the reverse process. The wavelength of light used is between 1300nm and 1500nm. To obtain such a wavelength, a laser chip is used to produce light at the required wavelength. After being emitted onto a 1-2mm^2 area, the light beam must be coupled with an optical fibre core whose diameter is 9 μm. To increase the efficiency of laser chip - optical fibre coupling, a lens system is also employed.

To transport the light along the optical fibre as far as possible, the maximum power must be transmitted. This requires that the light beam must match the characteristics of the optical fibre, in particular that the laser chip and the optical fibre must be precisely aligned. This alignment cannot be automatically made by a precision fitting since individual laser chips and optical fibres have different properties. The power transmitted decreases significantly with a change in alignment of only a few microns. Therefore, it is necessary to adjust the alignment dynamically. A human operator can carry out this task but it is a slow process. A computer-controlled system could be used, however this process is also slow because of the large number of measurements required to find the proper alignment. The matter is also complicated by the fact that the power does not follow a Gaussian distribution and there exist local maxima and noise. An efficient search algorithm able to find the global maximum of a multimodal search space is required to produce an optimal alignment in an appropriate time. In this work, simulated annealing is employed for this purpose.

4.1.2 Experimental Setup

The experimental system is shown in Figure 4.1. The mechanical unit consists of two platforms housing laser chip and optical fibre, respectively. The optical fibre platform is mobile and the laser chip platform is fixed. Three motors control the displacements, each with a range of 6mm. The alignment is carried out in three dimensions and the quality of alignment is determined through a measurement of optical power using an optical power meter. The three axes are oriented as shown in Figure 4.2. The x-axis is also called the focal axis.

To control the motors, the system employs a set of software functions. These include the programs to read the power, to find the displacements in three directions and to run the motors for the alignment. After the power is read, the simulated annealing algorithm is used to find the appropriate displacements in three directions and the motors are driven for the optimal alignment. A configuration (solution) is represented by a position of the optical fibre with regard to the laser chip.

Since the optimal alignment corresponds to the maximum transmitted light, the objective function to be minimised during optimisation is taken as the negative of the power measured at the end of the optical fibre. The cooling schedule employs a geometric type of temperature update rule. The temperature is decreased by a constant factor after a fixed number of configurations. The generation mechanism provides two different moves. One of the moves consists of perturbing all dimensions at the same time while the other creates a perturbation in only one dimension chosen randomly. A perturbation is made by adding a positive or negative value to the current position. Because of the precision of the motors, the perturbation cannot be continuous and cannot be less than 0.1 μm.

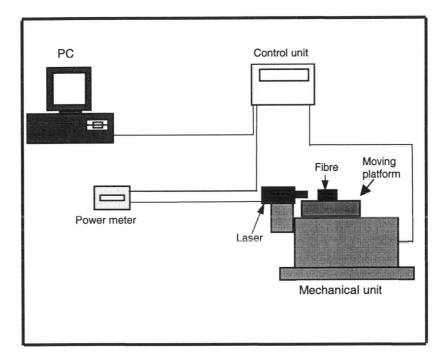

Fig. 4.1 Automatic alignment system

Since the fibre is moved in three directions, each direction has an impact on the alignment. However, the effect of change in each dimension on the alignment is different. A move along the y-axis or the z-axis leads to a significant change in the magnitude of the power, whereas a move along the x-axis has more influence on the distribution of the power. The global maximum increases when the fibre approaches the focal point and at the same time the local maxima decrease and a single global maximum is clearly observed.

As seen in Figures 4.3 and 4.4, the power is distributed in a small area around the global maximum, not larger than 100 μm² on the y - z plane. This is a very small area compared to the motor range of 6 mm. Moreover, moving along one of the axes does not imply the same change in the power as along the other axes. A move parallel to the y-axis or z-axis leads to a sharp decrease or increase of the power over a small distance (Figure 4.5), whereas a move parallel to the x-axis leads to a smaller change in power (Figure 4.6).

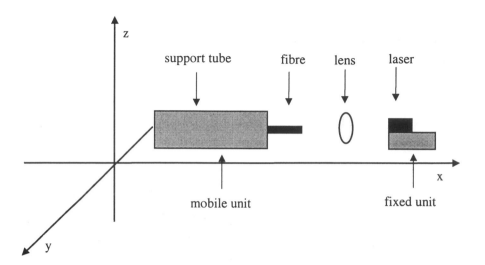

Fig. 4.2 Definition of axes of the alignment system

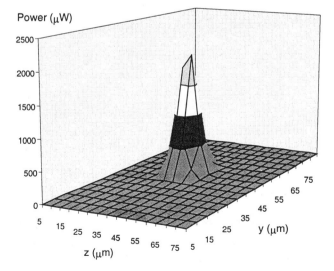

Fig. 4.3 Power distribution near the focal point

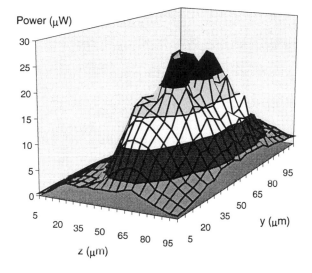

Fig. 4.4 Power distribution far from the focal point

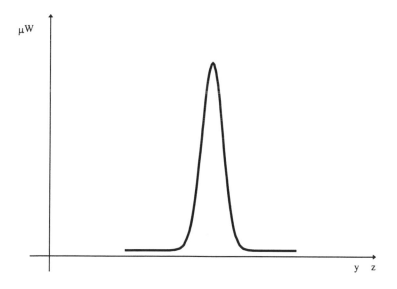

Fig. 4.5 Power distribution along y and z-axes

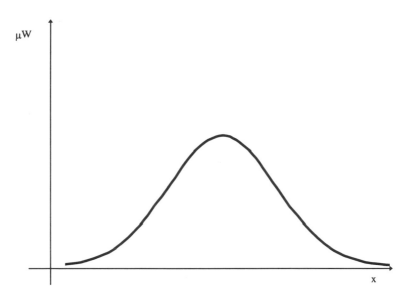

Fig. 4.6 Power distribution along x-axis

4.1.3 Initial Results

At the beginning of the experiments, the search space was not restricted and the initial solution was created randomly. In this case, it was observed that when the initial solution was chosen at a position where there is no power, it was very difficult for the algorithm to find the maximum point since there is a large area where the objective function has the same value.

When the search starts at a point where there is some power, the algorithm is able to find a maximum in the search space but not always the global maximum. That is because the algorithm is run for only a few seconds to obtain an alignment in a very short time, but within that period it cannot evaluate many solutions due to the relatively slow speed of the motors and it cannot complete even local convergence efficiently. The relation between the total number of trials carried out and the total time spent during alignment is shown in Table 4.1.

Table 4.1 Relation between the trial numbers and the time spent

Trials	40	50	100	245
Time	12s	15s	28s	65s

4.1.4 Modification of Generation Mechanism

A move in all directions lasts 3 times as long as a move in only one direction. Therefore, it seemed logical to encourage more one-dimension moves than three-dimension moves. In this case, the algorithm is able to try more moves during a single optimisation. However, both types of move are required for a successful search. The algorithm was therefore modified so that, on average, 40% of moves were three-dimension moves and the remaining 60% were one-dimension moves. It was observed that the results thus obtained were better than those in the initial experiments. The maximum was approached more often and more closely than before for the same number of moves.

These results also showed that the algorithm is able to find quickly a good maximum in the domain defined by axes y and z, but has a problem with optimising with respect to the x-axis. As explained before, this is due to the fact that the power is located in a small area around the global maximum in the search space defined by the y and z-axes but is more widely spread along the x-axis. Therefore, it was decided to generate more x-axis moves than other direction moves and to produce larger perturbations for that axis.

After modifying the generation mechanism again in this manner, the algorithm still had difficulty in finding the optimal x position. In particular, it was difficult to obtain higher values of x when the search started in a low x position. To help overcome this problem more positive perturbations in the x-axis were generated. The experiments were repeated and good results obtained when compared to the results of previous experiments. However, the global maximum was still not reached for every run. It was concluded that the starting solution had a crucial effect on the performance of the algorithm. A good starting solution must not have too small a power value and 0.2-0.3 μW seemed to be the minimum acceptable power to make a good start.

4.1.5 Modification of Cooling Schedule

The overall aim is to produce an optimal alignment as quickly as possible, say within 15-20 seconds. This allows the algorithm to evaluate 70 or 80 trials, including rejected and accepted solutions. The cooling schedule must be adjusted accordingly, and the 70 trials were divided into the number of temperature steps which was taken as 7.

Annealing was started with a temperature of 1, which was decreased by 10% after every 10 trials in 7 steps. Good results were obtained with this cooling schedule. It was observed that, during the first temperature steps, the algorithm accepts both good and bad solutions, but there was no significant change in the power. This is because at the beginning of a search there is not a great risk of falling into a local minimum trap, as local minima are close to the global minimum which is located far away initially. Towards the end of the search, the risk of being trapped at a local minimum is high and thus it will be more important to tolerate a temporary deterioration in the quality of the solution to escape from such a trap. In fact, the first two steps do not optimise greatly the alignment. The power begins to increase noticeably during the third temperature step. The temperature might be decreased very slowly, but in that case the search continues around the starting point and does not go towards a maximum until the third temperature step.

To overcome this drawback, a double annealing procedure was applied. The first annealing was designed to increase the power and move as close as possible to the area where the maxima are. During this part of the search the algorithm is unconcerned about locating the global maximum. The second annealing started at a higher temperature from the highest power value found by the first annealing procedure. This second annealing was designed to cause the algorithm to find the global maximum.

For the first annealing, five temperature steps were used and ten trials were evaluated for each step. The second annealing consisted of two temperature steps and ten trials were evaluated for each step. The algorithm with this double annealing reached the global maximum more often than that with the single annealing procedure. It was noted that the power value found by the first annealing was not far from the final power value.

When the double annealing procedure was employed, a problem arose due to the lack of precision of the motors. If the highest power value found by the first annealing is not the latest power value evaluated, the second annealing does not start with the desired maximum power, when the motors are returned to that highest power position. As an example, the original power reading may be 950µW while the power reading on return may be only 640µW.

4.1.6 Starting Point

All the runs were made with a fixed starting point determined manually. The starting point is critical to performance and is dependent on the particular laser transmitter component. To find a good starting point automatically where there is some power, some kind of initial scan is required. However, the use of this procedure significantly increases the time needed to complete the search. In the experiments, the automatic initial scan procedure, described below, took 5 to 100 seconds while the annealing took around 30 seconds to find the maximum.

The initial scan procedure consisted of scanning the search space in two different ways. From an initial position, a spiral scan is performed. The spiral scan is stopped when a power value larger than a threshold is found or when the scan process is too far away from the starting point. If sufficient power is not found the scan is carried out once more with a wider range and smaller steps. When the process stops at a position where there is some power, the algorithm starts to optimise this position in the y and z dimensions. For 4 steps, the algorithm searches for a larger power value through movement in the y and z-axes. Finally the remainder of the scan process is performed around the point that gives the best power value to find an improved value if one exists.

4.1.7 Final Modifications to the Algorithm

After all the above modifications, it was seen that the first annealing still had a problem in finding a good x position. To try and solve that problem, further modifications to the generation mechanism were implemented. 70% of perturbations were now made for the x-axis and 30% for both the y and z axes. Compared to the previous generation mechanism, the algorithm is forced to increase x values more often than decreasing them. As before, the second annealing is able to find more precise solutions since small perturbations are used in all directions. The fully automatic system with this annealing procedure was employed for the experiments. From the results obtained with several components it was seen that there was still a 20-25% failure rate to reach the maximum power. In the these cases, the search usually ended with too large a value for x and became trapped in a local maximum. The value of the x position could be increased to reach the global maximum. However, the laser beam is not parallel to the x-axis (focal axis). Thus, the (y, z) co-ordinates of the point of maximum power vary with different (y, z) planes at different locations along the x axis.

One way to overcome this problem is to scan the x-axis and to compare the best power value that can be found for each position. In other words, for each x position the algorithm has to optimise the alignment with regard to the y and z-axes. The algorithm changes the x position and re-optimises according to the plane. So the first part of the optimisation drives the search close to the area where the global maximum exists. After the first optimisation another annealing is performed to move closer to the global maximum.

A planar optimisation is performed to find a good global maximum according to the y and z-axes with an annealing algorithm in which the cooling schedule consists of three temperature steps. At each step, 5 solutions were evaluated. It was found that this number was sufficient to reach a good maximum. The time required to complete the planar optimisation was 4.5 seconds.

The focal axis can be scanned in two different ways. One way is sequential scanning. However, that is not the most efficient technique. For example, starting from x = −100 μm the range to x = 200 μm is scanned using a window width of 30 μm. This scan process will find the region effectively but the search space is restricted and the method time consuming. Another method, adopted here, is to move in the direction of increasing power values. When the power begins to decrease, the scan starts again from the higher value with a step which is ten times smaller than the previous step. The scanning stops when the step value reaches a threshold.

Another factor in the experiments was that the power sometimes reached the value of 2000 µW, which is the maximum power measurable by the power meter. In this case, the optimisation process was stopped as soon as the algorithm reached 2000 µW.

Table 4.2 Results obtained using simulated annealing algorithm and fuzzy logic control

Component number	Initial scan time	First step time	Second step time	Total time	Anneal- ing results	Fuzzy control time	Fuzzy control results
1	20	22	10	52	1103	49	980
2	25	49	11	85	1498	38	1472
3	29	54	3	86	1059	37	960
4	9	22	11	42	943	33	960
5	40	25	11	76	1816	45	1700
6	6	23	11	40	1415	23	1380
7	31	41	11	83	1235	57	1000
8	40	29	11	80	1137	46	1120
9	7	9	11	37	900	28	1000
10	8	19	8	35	980	29	980
11	6	6	0	12	2000	12	2000
12	23	23	11	57	1680	50	1680
13	15	36	9	60	1200	32	1080
14	12	5	0	17	2000	19	2000
15	18	12	0	30	2000	41	2000
16	5	2	0	7	2000	8	2000
17	24	32	12	68	1372	52	1230
18	17	29	8	54	1520	42	1390
19	9	16	11	36	1457	30	1460
20	10	30	11	51	1100	42	1100
21	8	34	6	48	1430	32	1200
22	38	53	11	102	1560	41	1100
23	9	10	0	19	2000	23	2000
24	22	13	0	35	2000	27	2000
25	18	34	11	63	1780	42	1800
26	6	11	0	17	2000	11	2000
27	6	7	0	13	2000	13	2000
28	9	23	11	43	1014	33	960
Average	**16.8**	**24.3**	**7.1**	**48.1**	**1507**	**33.4**	**1484**

4.1.8 Results

With the annealing algorithm having all these modifications, it always reached the global maximum when the planar optimisation was made correctly. In Table 4.2, the results obtained using the modified simulated annealing algorithm and, for comparison, a fuzzy control system are presented. In the table, time is in seconds and power in μW. The total annealing time includes the time to perform the initial scan, the focal axis scan time (first step) and the final annealing time (second step).

The power value of the global maximum found by the annealing algorithm is larger than the value obtained by the fuzzy logic controller. The difference between the power values produced using the two different methods is never larger than 200 μW. In terms of the average power, the simulated annealing algorithm shows marginally better performance than the fuzzy logic controller. However, the time spent by the annealing algorithm is longer than that of the fuzzy logic controller.

4.2 Inspection Stations Allocation and Sequencing

Most current manufacturing processes consist of a number of stages or production stations arranged in serial or complex non-serial configurations. Two important research topics in the area of multi-stage manufacturing are the design of inspection systems and the design of production inventory systems. The former is to ensure that the outgoing product quality level is consistent with the product specifications and the latter is to deal with the optimal level of lot-sizes at each production stage.

Multi-stage manufacturing systems, where the raw material is transformed into a final product through a series of production stages, present various possibilities for inspection. The quality of the outgoing product is clearly dependent on the quality of the product at each production stage. Inspection is carried out before or after a processing stage to ensure that only good quality products are passed to the next stage or leave the production line. Usually there are some important issues to be considered:

(a) Should each of the product units be inspected at particular production stages?
(b) If the answer to (a) is affirmative, 100% inspection is required. If it is negative, what is the optimal number to be inspected in a batch and what is the acceptance number?
(c) What is the best allocation and sequencing of the inspection stations?

There is a trade-off between inspecting too many units (which incurs high inspection costs and slows down the entire production line) and too few (which decreases the reliability of the output units). This implies two different optimisation problems namely, allocation and sequencing of inspection operations and design of optimal acceptance sampling inspection plans. In this section, an application of the simulated annealing (SA) algorithm to the problem of allocation and sequencing of inspection operations will be presented.

The next section (Section 4.3) describes the application of simulated annealing to the selection of optimum lot-sizes in multi-stage manufacturing systems, which is a problem related to batch-size optimisation with respect to each production stage.

4.2.1 Background

Inspection is concerned with separating product units that conform to specifications from those that do not and preventing non-conforming (defective) product units from reaching the customer or the external user.

The inspection activity can be carried out in many ways: manual (human inspectors), automated and hybrid, which is the combination of manual and automated systems. Recent technological developments in automated visual inspection, pattern recognition and image processing techniques have led to an increase in the implementation of automated systems. Errors and inconsistencies in manual inspection provide the motivation for this increased implementation.

However, higher implementation costs and technical difficulties can be associated with automated systems. The selection and location of such inspection stations must be carefully considered since it will have a significant effect not only on the product quality but also on the total cost of manufacturing.

The optimal allocation and sequencing of inspection stations is to be considered in this research. The notion of optimality in this case encompasses factors such as the cost of inspection, the cost of allowing a defective unit to be output, and the cost of internal failure (scrap and rework).

As mentioned before, a multi-stage manufacturing system can exist in a serial or non-serial configuration. Figure 4.7 shows these two types of systems. In a serial system, each processing stage except the first in the series has a single immediate predecessor. Also, each stage except the last has a single immediate successor. Inspection stations may exist between the processing stages. In a non-serial system, at a certain stage, the product may be assembled or joined with products from other processing lines. Hence the optimisation of the allocation of inspection stations is more complex.

Only a serial system is to be considered in this work on inspection stations allocation with the following general criteria [Raz and Kaspi, 1991]:

1. There are several production stages.
2. Discrete product units of a single type flow in a fixed linear sequence from one stage to the next.
3. Products flow in batches or lots of size one.
4. Each production stage consists of a single production operation followed by zero, one or more inspection operations in a fixed sequence.
5. Every production or inspection operation incurs a constant unit processing cost.
6. In every inspection operation there are two kinds of inspection errors:
    - classification of a conforming unit as non-conforming.
    - classification of a non-conforming unit as conforming.

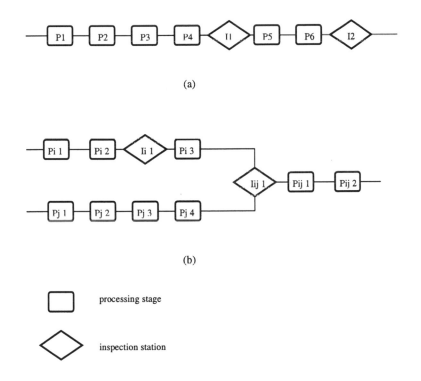

(a)

(b)

☐ processing stage

◇ inspection station

Fig. 4.7 (a) Serial production-inspection system   (b) Non-serial production-inspection system

These errors will impact on the cost, effectiveness and credibility of quality assurance efforts.

7. Product units classified as non-conforming are removed from the production line and are disposed of in two ways: scrap or rework.
8. Delivering a non-conforming unit to the customer will cause a penalty cost.

The optimal design of inspection stations has been studied since the 1960s. Raz [1986] surveyed the elements of the inspection stations allocation problems and the models that had been proposed in the literature. His paper quoted 17 models developed from 1964 to 1984 for serial and non-serial production systems. Most of the models used dynamic programming techniques. Ballou and Pazer [1982; 1985] developed a production-inspection model allowing for inspection errors in a serial production system. Chakravarty and Shtub [1987] suggested a shortest-path heuristic method to determine the strategic location of inspection activities and the production lot-sizes. Peters and Williams [1987] used dynamic programming and direct search techniques to determine the location of quality monitoring stations. The problem with dynamic programming is that when the number of processing stages increases the complexity of computation increases dramatically. A non-optimum solution would be accepted as optimum as the problem becomes larger. Furthermore, the optimisation decisions are taken separately, *stage by stage*, rather than by performing *global optimisation* of the multi-stage system.

More recent work has been carried out by Raz and Kaspi [1991]. The Branch-and-Bound technique was suggested to allocate imperfect inspection operations in a fixed serial multi-stage production system. This methodology still involves *stage by stage optimisation*. The application of Artificial Intelligence (AI) was suggested by Raz [1986]. Kang *et al.* [1990] proposed a rule-based technique to determine near-optimal inspection stations design. No result was presented in Kang's paper. The work reported here extends the work of Raz and Kaspi [1991] to *global optimisation* by the use of *AI* techniques.

4.2.2 Transfer Functions Model

The inspection stations allocation model is formulated with the objective of minimising the total cost per product unit. The total cost includes the unit cost of inspection and cost of defective items. There are two kinds of cost of defective items. One is the rework and replacement cost before the defective items are released from the company. The other is the penalty cost for each non-conforming unit reaching the customer.

The problem is to decide which inspection operations will be performed immediately following each production stage. The constraints on the optimisation problem are based on a required outgoing fraction of non-conforming units and the number of inspection operations.

In this work, the multi-stage production-inspection model developed by Raz and Kaspi [1991] is considered. The model is called the Transfer Functions Model (TFM). The TFM provides a unified framework for the analysis of multi-stage systems with different types of production and inspection operations. This model facilitates the implementation of the calculations required to find the optimal solution. The TFM at each stage, whether a production or inspection stage, is described by two transfer functions: the *Cost Transfer Function* and the *Quality Transfer Function*.

The *Cost Transfer Function* (CTF) relates the cumulative unit costs before and after completion of the production or inspection operation denoted by C_i and C_o respectively. The *Quality Transfer Function* (QTF) of an operation relates the probabilities q_i and q_o that a unit is non-conforming before and after the operation respectively. The TFM comprises the following parameters (see Figure 4.8):

δ: fraction of the units classified as non-conforming units that are scrapped. Therefore the fraction of the units reworked is $(1-\delta)$;

θ: probability that a non-conforming unit will be classified as conforming;

π: probability that a conforming unit becomes non-conforming as a result of improper performance of a production or repair operation;

ρ: probability that a non-conforming unit remains non-conforming after completion of a repair operation;

ϕ: probability that a conforming unit will be classified as non-conforming;

f: fraction of the cumulative cost of a scrapped unit that is retained as a salvage value;

PC: unit processing cost added to each unit;

RC: unit rework cost.

The final forms of both transfer functions after completion of the single operation are given as [Raz and Kaspi, 1991]:

$$q_o = \frac{(C_i + PC) - j(C_i + PC)\delta[q_i(1-\theta) + (1-q_i)\phi] + RC(1-\delta)[q_i(1-\theta) + (1-q_i)\phi]}{1 - \delta[q_i(1-\theta) + (1-q_i)\phi]}$$

$$\text{(4.1)}$$

$$q_o = \frac{q_i\theta + (1-\delta)\{\rho[q_i(1-\theta)] + \pi(1-q_i)\phi]\}}{1 - \delta[q_i(1-\theta) + (1-q_i)\phi]} \qquad \text{(4.2)}$$

These equations are applied to calculate the CTF and QTF for a single production or inspection station.

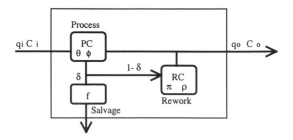

Fig. 4.8 Transfer functions model for a single processing and inspection stage

4.2.3 Problem Description

Consider a multi-stage manufacturing system, where there are up to 3 possible inspection stations to be located at any of 10 processing stages, stage 0 to stage 9. The raw material is input to stage 0 and the finished product is output from stage 9. The optimal location will minimise the total cost (See Figure 4.9).

The total cost $C_{T[s]}$ of a single stage is given in [Raz and Kaspi, 1991]:

$$C_{T[s]} = C_{o[s]} + q_{o[s]} \cdot PC_{[s+1]}$$

where $C_{T[s]}$ is the total cost at stage s and $PC_{[s+1]}$ is the production cost at the next stage $(s+1)$, $(s = 0$ to 9$)$.

The value of $C_{T[s]}$ depends on the inspection configuration at each stage.

The output of the CTF and QTF of one stage will be input to the next stage. Therefore the $C_{o[s]}$ and $q_{o[s]}$ are taken as $C_{i[s+1]}$ and $q_{i[s+1]}$. Equations 4.1 and 4.2 are applied again to find $C_{o[s+1]}$ and $q_{o[s+1]}$. The total cost for the next stage can then be described as:

$$C_{T[s+1]} = C_{o[s+1]} + q_{o[s+1]} \cdot PC_{[s+2]}$$

Stage 0:

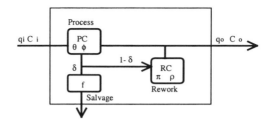

Stage 1:

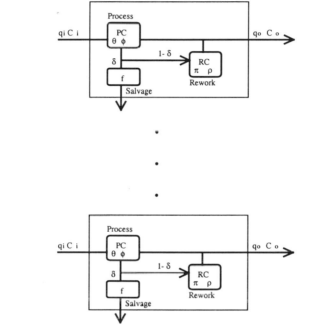

Stage 9:

Fig. 4.9 TFM for the whole production line

When the product exits the whole system and reaches the customer or external user, a penalty cost is incurred for each non-conforming unit. The unit penalty cost is PEN. The CTF is denoted by $C_{T[p]}$:

$$C_{T[p]} = C_{o[9]} + q_{o[9]}.\text{PEN}$$

The total cost per product unit, C_T, for the entire system can be described as:

$$C_T = C_{T[9]} + C_{T[p]} \qquad\qquad (4.3)$$

C_T is the objective function of the inspection stations allocation and sequencing problem. The values of the production and inspection parameters (π, δ, ρ, etc.) used in this work are given in [Hassan, 1997].

4.2.4 Application of Simulated Annealing

Representation. There are 10 processing stages with 3 types of inspection station, $I1$, $I2$ and $I3$. The processing stages are arranged serially. The inspection stations can be allocated to any stage of processing. There will be 16 possible inspection configurations at any single stage (see Figure 4.10), and the total number of configurations for all processing stages is 16^{10} (2^{40} possible solutions). This is because there are 10 processing stages and at each stage there are $\sum_{i=0}^{3} \dfrac{3!}{(3-i)!}$ or 16 possible inspection arrangements. (i is the number of possible inspection stations, see also Figure 4.10).

The problem was represented with a 40-bit binary string. This representation is shown in Figure in 4.11 and in Table 4.3. The motivation to apply the binary representation was its flexibility and ease of computation.

Re-configuration Mechanism. *Inversion* and *mutation* were applied as re-configuration operators (see Chapter 1).

Cost Function. The cost function for this problem is the objective function given in the previous section.

Cooling Schedule. The annealing process started at a high temperature, $T = 600$ units, and so most of the moves were accepted. The cooling schedule was represented by:

$$T_{i+1} = \lambda T_i$$

where λ is the cooling rate parameter, which was determined experimentally. Three cooling rates were used ($\lambda = 0.9, 0.98, 0.99$). With $\lambda=0.9$, the temperature reduced rapidly. With λ set to 0.98, the cooling rate was medium. For $\lambda=0.99$, the rate of cooling was very slow.

The initial stopping criterion was set at a total unit cost of 265.38. This value was the best total cost unit found for the same problem when a genetic algorithm (GA) was employed. A lower value was subsequently implemented [Hassan, 1997].

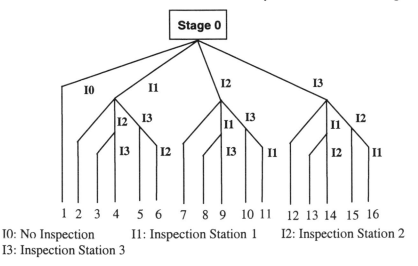

I0: No Inspection   I1: Inspection Station 1   I2: Inspection Station 2
I3: Inspection Station 3

Fig. 4.10 Possible inspection stations configurations in a single stage

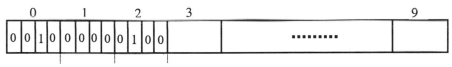

Fig. 4.11 Binary representation for 10 processing stages

Table 4.3 Binary representation of inspection stations in a single stage

Binary	Decimal	Order of
0000	0	No inspection
0001	1	I1
0010	2	I1-I2
0011	3	I1-I2-I3
0100	4	I1-I3
0101	5	I1-I3-I2
0110	6	I2
0111	7	I2-I1
1000	8	I2-I1-I3
1001	9	I2-I3
1010	10	I2-I3-I1
1011	11	I3
1100	12	I3-I1
1101	13	I3-I1-I2
1110	14	I3-I2
1111	15	I3-I2-I1

4.2.5 Experimentation and Results

Figure 4.12 (a-c) shows the optimisation curves for 3 experiments with different values of λ (0.9, 098, 0.99). Experiment 1 showed very random perturbation at the beginning of the iterations. The cost then quickly reduced to a local optimum. The programme was stopped after 12 hours. The total unit cost reached was 275.40. Experiment 2 showed much random perturbation at the beginning. During the first quarter of the optimisation process, there was improvement but the cost was very unstable. This perturbation was necessary to avoid local optima. Then the cost stabilised. When the temperature was low, the total unit cost began to decrease, until it reached the stopping criterion, which is the expected optimal solution to this problem. The algorithm terminated at 136 iterations after 1 hour 38 minutes of computation time. Experiment 3 showed very much random perturbation until 3/4 of the maximum number of iterations (200 iterations). Then, the cost started to stabilise. However before attaining the stopping cost value, the algorithm reached the maximum number of iterations with a total unit cost of 274.06. The maximum number of iterations could have been increased, but this would have been computationally expensive.

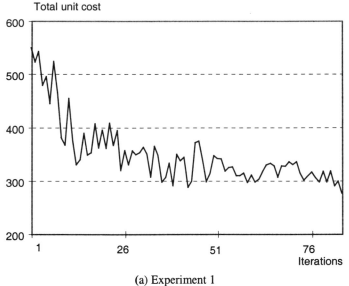

(a) Experiment 1

Fig. 4.12 Optimisation curve for inspection stations allocation

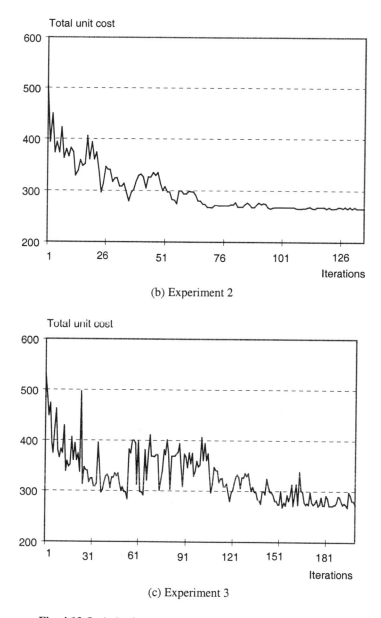

(b) Experiment 2

(c) Experiment 3

Fig. 4.12 Optimisation curve for inspection stations allocation

Experiment 2 was conducted again with a lowered stopping criterion. However no improvement was found after a very long computation time (12 hours). Therefore the solution found by the genetic algorithm was accepted as the optimal solution. The final inspection configuration is shown in Table 4.4.

Table 4.4 Final result for inspection stations allocation by *SA*

Stage	Binary no.			
0	0	0	0	0
1	0	0	0	0
2	0	0	0	0
3	0	0	0	0
4	0	0	0	0
5	0	0	0	0
6	0	0	0	0
7	0	1	1	0
8	0	0	0	0
9	0	0	1	0

Order of Inspection Stations: I2 after Stage 7; I1 followed by I2 after Stage 9.

Table 4.5 shows the best results found by the *SA* procedure and, for comparison, by the Branch-and-Bound technique [Raz and Kaspi, 1991].

The results of the experiment have confirmed that the initial temperature and cooling rate determine the quality of the solutions. If the initial temperature is too high and the cooling rate is too low, the configuration cannot achieve the optimal solution before it reaches the maximum number of iterations. If the cooling rate is too high, the process could become lodged in a local optimum. It was also noted that increasing the number of iterations did not have much effect on the final solution.

Table 4.5 Comparison between *SA* and the Branch-and-Bound Technique for inspection stations allocation and sequencing (Hassan, 1997)

Allocation and Sequencing	Simulated Annealing	Branch and Bound
Experiment 2	265.38	267.52

4.3 Economic Lot-Size Production

The need to compete and succeed has led to a number of technological developments in manufacturing systems. These include Flexible Manufacturing Systems (FMSs), Optimised Production Technology (OPT) and Computer Integrated Manufacturing (CIM). Recent developments in the area are Just-In-Time (JIT), Total Quality Management (TQM) and Total Preventative Maintenance (TPM). The on-going concern is to provide good quality products and to improve profitability via set-up cost reduction, zero defects, and reduction of manufacturing cycle times. A related issue of concern is achieving smaller lot-size production. This is possible only when set-up costs are reduced. Also, there exists a strong relationship between batch size and inventory cost in dynamic process control [Gunasekaran *et al.*, 1995]. For example, set-up cost reduction facilitates a reduction in inventory cost due to dynamic process quality control using smaller batch sizes for processing the items [Gunasekaran *et al.*, 1995].

Research into economic lot-sizes in multi-stage manufacturing systems is relatively new compared to other optimisation problems. Szendrovits [1975] presented a mathematical model for determining the manufacturing cycle time as a function of the lot-size in a multi-stage production inventory system. This model considers the cost due to the work-in-process inventory and establishes a relation between lead time and inventory cost. Goyal [1978] extended this work by optimising simultaneously both the lot-sizes and number of sub-batches. This results in a significant saving in total inventory cost. However, both of the models consider uniform batch sizes for all stages. Imo and Das [1983] studied the scheduling of batch-sizes for various production stages. Different batch-sizes with respect to each stage were considered, but work-in-process inventory cost was not taken into account. Karmarkar *et al.* [1995] studied the effect of work-in-process inventory cost due to queuing of batches. Porteus [1985] studied the economic trade-offs associated with very small lot-size production systems. The work considered the option of using set-up cost reduction in the classical undiscounted Economic Order Quantity (EOQ) model and determined an optimal set-up cost level. Porteus [1986] demonstrated that lower set-up costs and associated reduced lot-sizes can benefit production systems by improving quality control. Porteus's model was developed to show significant relationships between quality and lot-size. The model demonstrated that while producing a lot, the process can go *out of control* with a given *probability*. Once out of control, the process produces defective units until the end of production. The system incurs an extra cost for rework and related operations for each defective unit it produces. Thus, there is an incentive to produce smaller lots and have a smaller fraction of defective units.

However, there is no model to explain the relationship between quality control and set-up reduction in a multi-stage manufacturing system considering dynamic process quality control. Gunasekaran *et al.* [1995] studied this relationship and proposed a mathematical model. The model determined the EOQ and the optimal investments in set-up cost and process drift rate reduction programmes to minimise total cost. Gunasekaran's model is used in this work.

4.3.1 Economic Lot-Size Production Model

Consider a manufacturing system which consists of a number of production stages and products. The basic objective is to employ the optimal lot-size that will minimise the total cost of manufacturing. The mathematical model developed by Gunasekaran *et al.* [1995] will be considered. The model consists of 6 main cost functions per unit time.

1. *Cost of investment in set-up cost reduction programme, C_1*

$$C_1 = \sum_{i=1}^{M} \sum_{j=1}^{N} U_{ij} \tag{4.4}$$

where U_{ij} is the discounted cost of investment in a set-up reduction programme per unit time *($/year)*, i is the product number *($i=1$ to M)*, and j is the stage number *($j=1$ to N)*.

2. *Set-up Cost, C_2*

The total set-up cost considering all products and stages is given by:

$$C_2 = \sum_{i=1}^{M} \sum_{j=1}^{N} \left\{ \frac{D_i}{Q_{ij}} \right\} A_{ij} \tag{4.5}$$

where D_i is the demand for product i per unit time *(units/year)*, Q_{ij} is the batch size for product i at stage j; lower bound $d_{ij} \le Q_{ij} \le$ upper bound u_{ij} , and A_{ij} is the set-up cost per set-up for product i at stage j *($)* and is given as:

$$A_{ij} = \frac{K_{ij}}{U_{ij}^{e_{ij}^u}}$$

where K_{ij} is a positive constant (defined as the set-up cost per set-up when the elasticity of investment in the set-up cost reduction programme, e^u_{ij}, is zero).

3. *Cost of investment in process drift rate reduction programme, C_3*

$$C_3 = \sum_{i=1}^{M} \sum_{j=1}^{N} V_{ij} \qquad (4.6)$$

Where V_{ij} is the discounted cost of process control improvement methods *($/year)*.

4. *Cost due to process control, C_4*

$$C_4 = \sum_{i=1}^{M} \sum_{j=1}^{N} \left[\frac{D_i t_{ij} \alpha_{ij}}{\beta_{ij}} \right] y \qquad (4.7)$$

where t_{ij} is the processing time per unit of product i at stage j *(years/unit)*, β_{ij} is the mean service rate for bringing the process to normal operating condition for product i at stage j, and y is the cost per unit time of controlling the process activities *($/year)*. α_{ij} is the mean process drift rate while processing product i at stage j and is given as:

$$\alpha_{ij} = \frac{L_{ij}}{V_{ij}^{e_{ij}^v}}$$

where L_{ij} is a positive constant (defined as the process drift rate when the elasticity of investment in the process drift rate reduction programme, $e^v{}_{ij}$, is zero).

5. *Inventory cost due to process control and operations, C_5*

$$C_5 = \sum_{i=1}^{M} \sum_{j=1}^{N} \left[D_i T_{ij} G_{ij} \{ \frac{\alpha_{ij}}{\beta_{ij} - \alpha_{ij}} \} \right] H \qquad (4.8)$$

where T_{ij} is the processing time for a batch of product i at stage j *(years)*, given by $T_{ij} = Q_{ij} t_{ij}$. G_{ij} is the mean cost per unit of product i between stages j and $j+1$ *($/unit)* and is given by:

$$G_{ij} = (C_{ij} + C_{i[j+1]})/2$$

where C_{ij} is the cost per unit product i after processing at stage j *($/unit)* and H is the inventory cost per unit investment per unit time period *($/$ year)*.

6. *Inventory cost due to queuing of batches,* C_6

$$C_6 = \sum_{i=1}^{M} \sum_{j=1}^{N} \left[D_i C_{ij} \frac{\lambda_{ij}}{\lambda_{i[j+1]}(\lambda_{i[j+1]} - \lambda_{ij})} \right] H \tag{4.9}$$

where λ_{ij} is the production rate for product i at stage j; $\lambda_{ij} = \dfrac{x_{ij} \times S_j}{F_{ij}}$, x_{ij} is the

priority assigned in processing product i at stage j and F_{ij} is the mean completion

time for a batch of product i at stage j *(years)* given by $F_{ij} = T_{ij} \left[\dfrac{\beta_{ij}}{\beta_{ij} - \alpha_{ij}} \right]$, and S_j

is number of machines at stage j.

The total cost, C_{LT}, is the sum of the costs C_1 to C_6.

The problem of economic lot-size production addressed here can be formulated as follows. There are three stages of production, each manufacturing three different products. The objective is to determine optimal batch sizes, Q_{ij}, optimal annual costs of investment in set-up, U_{ij}, and optimal annual cost of process control improvements, V_{ij}, to minimise the total cost, CLT. The values of the parameters adopted in this work are given in [Hassan, 1997]. The objective function for this problem is the total cost, CLT.

Assumptions. The assumptions made are as follows:

1. Demand per unit time of the product is deterministic and known;
2. Set-up cost per set-up is constant and independent of set-up sequence and batch sizes;
3. Process drifts follow a Poisson distribution and the service time required for each drift follows an exponential distribution with mean drift and service rates, respectively. The mean service rate β_{ij} is higher than the mean process drift α_{ij};
4. Machines at each stage have identical capacities;
5. There is no finished product inventory as the products will be dispatched once processing is completed at the final stage;
6. There are no machine breakdowns;
7. Cost per unit service time required to bring the process to normal condition is the same for all the products and at all stages is known;
8. Investment in process drift rate and set-up cost reduction programmes leads to favourable results;
9. Elasticities of the cost of investment in set-up cost and process drift-rate reduction programmes per unit time are constant and known.

4.3.2 Implementation to Economic Lot-Size Production

The main task is to determine the optimum values of batch sizes Q_{ij}, *investment* U_{ij}, and V_{ij} in set-up cost and process drift rate reduction programmes, and hence minimise the total cost of manufacturing. From the problem definition there are 27 parameters (9 for each Q_{ij}, U_{ij}, and V_{ij}; i=1 to 3, j=1 to 3) to be determined.

Implementation Details. A binary string is used to represent the 27 parameters. Q_{ij} is an integer value ranging between 100 and 600 and is represented by 9 bits. U_{ij} and V_{ij} are real values ranging between 150.00 and 400.00 represented by 11 bits. The total size of the solution space for this problem is 2^{279} as the total string length is 279 bits (9x(9+11+11)). Again, *mutation* and *inversion* were used as reconfiguration techniques to create new solutions.

The objective function E was the cost function C_{LT} as described in the previous section. The control parameters, i.e. the initial temperature T and the cooling rate λ, were again empirically determined.

Experimentation and Results. Experiments were carried out with 4 different sets of control parameters. Experiment 1 was conducted at a relatively high initial temperature (10 *units*) and slow cooling rate (0.98). In Experiment 2, the same initial temperature was set, with a higher cooling rate (0.95). In Experiment 3, a low initial temperature (*1 unit*) and slow cooling rate (0.98) were chosen. In Experiment 4, a low initial temperature (*1 unit*) and faster cooling rate (0.95) were used. The stopping criterion was set at C_{LT} = 7667, which was the lowest cost found when a genetic algorithm was applied to the same problem. Another stopping criterion was the computation time, which was set at a maximum of 12 hours. Finally, the third criterion was the number of iterations, limited to 300.

The optimisation curves for each experiment are shown in Figure 4.13 (a-d). The final cost value found in Experiment 1 was high (C_{LT} = 8047). The change in total cost per iteration was still high when the algorithm reached the maximum number of iterations. The solution is far from optimal. For Experiment 2, the probability of acceptance decreased rapidly (rapid cooling). The optimisation was then lodged in a local optimum. The computation time increased from one iteration to another. The algorithm stopped at 143 iterations after reaching the maximum computation time. The final cost value was 7942.

In experiment 3, a low initial temperature was used with a slow cooling rate. It was expected that the problem of high perturbation and local optima of previous experiments would be solved. From the optimisation curves, the improvement of the performance is confirmed. The algorithm started with high perturbation but this gradually reduced until it was almost zero. The minimum cost found from this experiment was C_{LT}= 7690 where the process stopped after reaching the maximum number of iterations.

In experiment 4, the optimisation was abandoned at a local optimum. The experiment was stopped at iteration 196 after 12 hours of computation. The lowest total cost found was C_{LT} = 7846.

This series of experiments again show the importance of selecting the *SA* control parameters correctly. Experiment 3, where a low initial temperature and slow cooling rate were adopted, produced the best result for the current problem. Table 4.6 presents the best results found by *SA* and, for comparison, by the Direct-Search method of Gunasekaran *et al.* [Gunasekaran *et al.* 1995]. Table 4.7 lists the values of the 27 parameters found in Experiment 3.

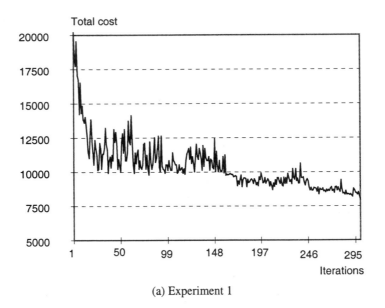

(a) Experiment 1

Fig. 4.13 Optimisation curves for economic lot-size production

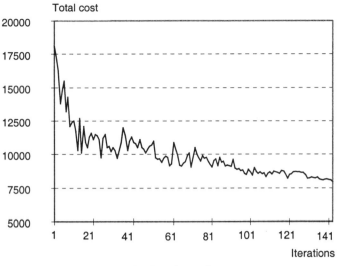

(b) Experiment 2

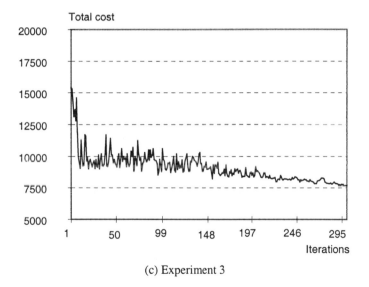

(c) Experiment 3

Fig. 4.13 Optimisation curves for economic lot-size production

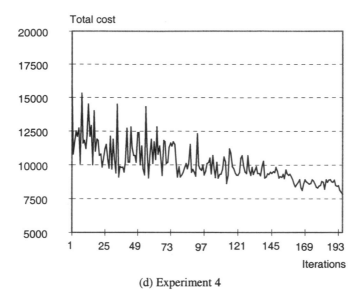

(d) Experiment 4

Fig. 4.13 Optimisation curves for economic lot-size production

Table 4.6 Results of *SA* and direct search for economic lot-size production problem

Lot-Size Production	Simulated Annealing	Direct Search
Experiment 3	7690	7947

Table 4.7 Detailed results for Experiment 3

	Stage		Product 1	Product 2	Product 3	Total Cost, C_{LT}
		1	373	369	194	
	Q	2	539	364	366	
		3	550	148	232	
		1	286.07	239.37	285.10	
Expt. 3	U	2	229.60	251.58	200.39	7690.381
		3	206.25	237.61	208.89	
		1	360.43	249.49	236.37	
	V	2	261.33	225.39	201.71	
		3	360.47	200.47	298.99	

4.4 Summary

This chapter has described three engineering applications of simulated annealing. The first is the optical fibre and laser alignment problem. The second and third applications are inspection stations allocation and sequencing, and optimal lot-size production. It can be seen in all three cases that the results obtained compare well with those produced by more conventional optimisation methods.

References

Ballou, P.B. and Pazer, L.H. (1982) The impact of inspection fallibility on the inspection policy in serial production systems, *Management Science*, Vol.28, No.4, pp.387-399.

Ballou, P.B. and Pazer, L.H. (1985) Process improvement versus enhanced inspection in optimised systems, *Int. J. of Production Research*, Vol.23, No.6, pp.1233-1245.

Barrere, M. (1997) *Simulated Annealing to Solve Optimisation Problems*, Internal Report, Intelligent Systems Research Laboratory, School of Engineering, Cardiff University, UK.

Chakravarty, A.K. and Shtub, A. (1987) Strategic allocation of inspection effort in a serial, multi-product production system, *IIE Transactions*, Vol.3, pp.13-22.

Goyal, S.K. (1978) Economic batch quantity in a multi-stage production system, *Int. J. of Production Research*, Vol.16, No.4, pp.267-273.

Gunasekaran, A., Korukonda, A.R. , Virtanen, I. and Yli-Olli, P. (1995) Optimal investment and lot-sizing policies for improved productivity and quality, *Int. J. of Production Research*, Vol.33, No.1, pp.261-278.

Hassan, A.B. (1997) *Intelligent Optimisation Techniques for Multi-Stage Manufacturing Systems*, PhD Thesis, School of Engineering, Cardiff University, UK.

Imo, I.I. and Das, D. (1983) Multi-stage, multi-facility, batch production system with deterministic demand over a finite time horizon, *Int. J. of Production Research*, Vol.21, No.4, pp.587-596.

Kang, K.S., Ebeling, K.A. and La, S.H. (1990) The optimal location of inspection stations using a rule-based methodology, **Computers Ind. Engineering**, Vol.19, No.1-4, pp. 272-275.

Karmarkar, U.S., Kekre, Sh. and Kekre, Su. (1995) Lotsizing in multi-machine job shops, **IIE Transactions**, Vol.17, No. 3, pp. 290-298.

Peters, M.H. and Williams, W.W (1987) Economic design of quality monitoring efforts for multi-stage production systems, **IIE Transactions**, Vol.3, pp.85-87.

Porteus, E.L. (1985) Investing in reduced setups in the EOQ model, **Management Science**, Vol.31, No.8, pp.998-1010.

Porteus, E.L. (1986) Optimal lot sizing, process quality improvement and setup cost reduction, **Operations Research**, Vol.34, No.1, pp.137-144.

Raz, T. (1986) A survey of models for allocating inspection effort in multistage production systems, **J. of Quality Technology**, Vol.18, No.4, pp.239-247.

Raz, T. and Kaspi M. (1991) Location and sequencing of imperfect inspections in serial multistage production systems, **Int. J. of Production Research**, Vol.29, No.8, pp.1645-1659.

Szendrovits, A.Z. (1975) Manufacturing cycle-time determination for a multi-stage economic production quantity model, **Management Science**, Vol.22, No.3, pp.298-307.

Chapter 5

Neural Networks

This chapter consists of two main sections. The first describes the application of mapping and hierarchical self-organising neural networks (MHSOs) for Very-Large-Scale-Integrated (VLSI) circuit placement [Zhang and Mlynski, 1997]. The second presents the application of the Hopfield neural network to the satellite broadcast scheduling problem [Funabiki and Nishikawa, 1997].

5.1 VLSI Placement using MHSO Networks

An important step during VLSI circuit design is to determine the locations of the circuit modules. This requires an optimisation process in which criteria such as minimum chip area and minimum wire length must be satisfied. Although there are several techniques in the literature proposed for placement problems [Qinn, 1975; Forbes, 1987; Breuer, 1977; Sait and Youssef, 1995; Sechen and Sangiovanni-Vincentelli, 1985; Shahookar and Mazunder, 1990; Casotto and Sangiovanni-Vincentelli, 1987; Kling and Banerjee, 1989] it still remains an important task to develop a powerful technique which is able to find an optimal solution in a reasonable computation time. Several researchers have attempted to develop a successful technique based on neural networks. The Hopfield network has been applied in [Yu, 1989]. However, the results obtained showed that the method requires long computation time and the parameters of the network are too sensitive to control [Zhang and Mlynski, 1997]. More promising results have been reported in [Zhang and Mlynski, 1997] for the application of modified Kohonen self-organising neural networks to this problem and the remainder of this section describes this work.

The Kohonen self-organising neural network is based on the neural somatotopical mapping [Kohonen, 1989] and has been successfully used for the travelling salesman optimisation problem [Angeniol et al., 1988]. Since the two-dimensional implementation of the mapping is very similar to the two-dimensional placement problem, Kohonen's network has also been applied to the circuit placement problem by several researchers [Hemani and Postula, 1990; Rao and Patnaik, 1992]. However, in the implementation, if the neurons of the neural network are directly replaced with circuit modules, some undesirable problems occur [Zhang and Mlynski, 1997]. Therefore, Zhang and Mlynski [1997] proposed two versions of mapping and hierarchical self-organising network (MHSO), which do not suffer from these problems.

In the new approach proposed by Zhang and Mlynski [1997], the characteristics of the placement problem are combined with the mapping property of a neural model. The approach is based on the somatotopical mapping, and topology maps between the input space and the output space are created by the algorithm. The input space is the somatosensory source and the output space is a space where the circuit modules are located. In the learning phase, the slots that exist in the placement carrier in the input space are "stimulated" randomly and the "most responsible" circuit module in the output space is selected competitively. The neural network learns from a sequence of movements of circuit modules. During the learning process, the network allows relevant modules to cooperate and gradually organises itself toward an optimal solution. The learning process is self-organising without supervision [Zhang and Mlynski, 1997]. The algorithm was first designed for the grid placement problem. In this problem, slots exist in the placement carrier (in gate arrays) or for positions within columns in standard cell circuits. This version of the algorithm is called MHSO.

The MHSO algorithm was then extended to deal with the gridless placement problem. In this case, there are no predefined slots, as in the case of macro cell placement. This version of the algorithm is called MHSO2 and operates in a hierarchically organised manner. A neural network is employed to describe the connection wires of the circuit modules. To produce the final placement of macro cells with arbitrary shapes and sizes, the shape and the dimension of each module are simultaneously considered by means of subordinate neural networks.

The common features of both algorithms are summarised as follows. First, there is no requirement for the definition of an objective function. Second, the final result to be obtained does not depend on the initial placement. Third, the data structure of the algorithms is straightforward since no matrices are used. Finally, modules are directly mapped into the predefined slots on the placement carrier and the shape and the size of modules are considered along with the total wire length by using hierarchical neural placement. The orientations and rotations of modules are also considered in the macro cell placement. Zhang and Mlynski [1997] obtained successful results for gate arrays and standard cell circuits using the MHSO algorithm, and for macro cell circuits using the MHSO2 algorithm. In the following sub-sections, first the placement model based on mapping self-organising networks proposed by Zhang and Mlynski [1997] and the hierarchical neural network will be described, and then the results obtained will be summarised.

5.1.1 Placement System Based on Mapping Self-Organising Network

In the neural placement system, the coordinates of the circuit modules in two-dimensional output space are represented by the weight vectors. The modules are distributed in the output space and identified by their positions w. Here, w represents the mapping vector. A change of the vector corresponds to a movement of modules. Each vector x_j ($j=1,2,...,m$) in the input space represents the position of a slot which is available for module placement. m is equal to the number of slot positions.

For a set of modules $M=\{1,2,...,i,..,n\}$, a somatotopical mapping $\varnothing_p$ is defined for placement p by the mapping from the placement carrier in the input space E onto the modules in the output space A:

$$\varnothing_p : E \to A, \quad x \in E \to i \in A \tag{5.1}$$

The "most excited" module e ($e \in M$) is chosen according to the following condition:

$$\|x-w_e\| = \min_{i} \in A \|x-w_i\| \tag{5.2}$$

where w_e and w_i are the mapping vectors which represent the positions of modules e and i, respectively. Figure 5.1 shows the input and the output spaces of the placement system.

During the learning process, the modules are excited one after another by stimuli given in the input space. The groups of the excited modules that give the influencing region of the module e are determined through connections to a module e, which is excited.

The spatial neighbourhood of neurons in the system must be mapped to the topological neighbourhood of modules in a circuit. The module set M consists of subsets named layers $S_k \subset M$, such that:

- layer S_{k+1} contains all modules that connect with modules at S_k and are not included in S_i ($i=1,...,k$) layers (in other words, all layers are disjunct),

- $M = \bigcup_{k=1}^{u} S_k$.

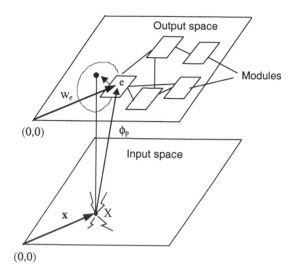

Fig. 5.1 Neural placement system

In the neural placement system, the functional neighbourhood is used to consider the connectivity between modules. The functional neighbourhood that corresponds to the connectivity relationship is defined as a connection zone $Z_e(s)$. This is the region influenced by a module e at depth s. The zone $Z_e(s)$ is composed of the modules in the s layers:

$$Z_e(s) = \bigcup_{k=1}^{s} S_k \tag{5.3}$$

In fact, $Z_e(s)$ plays a cooperation role. The range of cooperation reduces as time t increases. The range of the connection zone is initially selected to be large so that all modules are covered. The depth s decreases as time continues and thus the range of maximum activity becomes localised. Following convergence, the zone $Z_e(1)_{t=tmax}=S_1=\{e\}$ and $s=1$.

The excitation value of the modules at layer S_k depends on both layer index k and the momentary depth $s(t)$ of the connection zone:

$$L(k) = \exp(-\frac{c.k}{s(t)}) \tag{5.4}$$

where c is a constant.

When a module j falls into a connection zone $Z_e(s)$, the function is trapeze-shaped. Otherwise, it is zero outside the region. It is clear that $L(k)$ will be largest if $j=e$ and will reduce to zero with increasing layers.

The following model for learning rate is used:

$$H_i = L(k)V_i \tag{5.5}$$

where V_i is the connectivity weight, also called connection strength, of module i to module j that exists at the higher layer. The connection strength is computed by:

$$V_i = \max_j(\sum_k(\frac{n_k + b}{n_k})) \tag{5.6}$$

where $k \in N(i) \cap N(j)$, and $N(i)$ and $N(j)$ are the wires associated with the pins of the modules i and j, n_k is the number of pins in connection k and b is the weight parameter which is usually equal to 1. In the simplest way, V_i is equal to the number of the connections between the module i and the linked module at a higher layer.

Hence, the adaptation of the mapping vector w_i corresponding to input x is defined as:

$$
\begin{aligned}
w_i^{(t+1)} &= w_i^{(t)} + \varepsilon H_i(x - w_i^{(t)}) \qquad \forall i \in Z_e(s(t)) \\
&\quad w_i^{(t)} \qquad\qquad\qquad\qquad \forall i \notin Z_e(s(t))
\end{aligned}
\tag{5.7}
$$

where x is the two-dimensional vector of the input space and ε scales the overall size.

Using this learning rule, all modules existing in the connection zone $Z_e(s)$ move in the direction of x and their positions are modified step by step. In the placement system, the vectors w_i store the knowledge of the placement, while the up-to-date knowledge is provided through the stimuli.

The operation of the placement system is described as follows:

1. Initial placement: Start with initial values chosen randomly for w_i, i.e., arbitrarily distributed positions of modules.
2. Stimulus: Choose a random vector x to represent a position on the placement carrier according to the probability density $p(x)$.
3. Response: Determine the excited module e based on equation (5.2) and compute the relevant connection zone $Z_e(s)$.
4. Learning: Change the vector w_i according to equation (5.7) and go to step 2.

In the self-organising mapping algorithm, the conversion from similarity of stimulating signals into a neighbourhood of excited neurons constitutes the optimisation process. Hence, the neurons of the network that have similar tasks can communicate with each other because of extremely short connections. The similarity relations are transformed into connection relations among modules within a functional neighbourhood in the placement algorithm.

In the use of the MHSO algorithm for gate arrays, the probability distribution $p(x,y)$ concentrates on a set of randomly chosen slot positions. One of the inputs x_1, x_2, ..., x_j, ..., x_m is presented at time t. The probability distribution of the "touch stimuli" is changed according to the design style. A random number generator is used to obtain the points of x which are supposed to be distributed over an array that contains the slot information:

$$p(x,y) = \frac{1}{2\pi\sigma_x\sigma_y}\exp[-\frac{1}{2}\{(\frac{x-\mu_x}{\sigma_x})^2 + (\frac{y-\mu_y}{\sigma_y})^2\}] \tag{5.8}$$

In the use of the MHSO algorithm for standard cell circuits, Gaussian centering is employed on the middle of cell columns where there is higher probability than in the channel region used for routing. The probability distribution is:

$$p(x) = \frac{1}{M}\sum_{i=1}^{M}\frac{1}{\sqrt{2\pi}\sigma}\exp(-\frac{(x-a_i)^2}{2\sigma^2}) \tag{5.9}$$

where $\sigma > 0$ and a_i is the x-coordinate of the middle axis of cell column i. M denotes the number of columns.

The chip area occupied by the standard cells is constant. This area is equal to the sum of cell areas. The placement can be optimised to minimise the area occupied by the interconnected wiring and the other unoccupied areas outside the cells. Also, the area required for interconnected wiring is only directly affected by the vertical segments of the inter-connections, while the routing channel's width change Δb is not affected by the horizontal segments. Therefore, a goodness G estimation of vertical wire segments in the routing channels is expressed as:

$$G = \frac{1}{k}\sum_{i=1}^{k}\frac{V_{xi}}{V_{xi}+V_{yi}} \tag{5.10}$$

where V_{xi}, V_{yi} are the total length of horizontal and vertical wire segments in routing channel i and k represents the number of routing channels.

It is clear from equation (5.10) that, to increase the goodness V_{yi} must be reduced. To do that, a new metric is introduced. In the new metric, the ratio of the vertical distance a between two positions in the same column to the horizontal distance b from the position in the neighbouring column is not predetermined at 1:1. The vertical distance is taken to be r times the horizontal distance between the cells

$$a = r.b \tag{5.11}$$

With the aid of this metric most connections are made horizontally so that results are reached in accordance with the predefined goodness.

5.1.2 Hierarchical Neural Network for Macro Cell Placement

The MHSO algorithm presented in the previous sub-section deals with standard cell circuits. This section describes a version of MHSO, MHSO2, for placement problems where the modules are of different sizes and shapes, for example, the macro cells.

The neural placement system is extended by forming a new set of neural networks, each of which, respectively, represents an approximate module area. This new version of the algorithm is based on a hierarchical neural network and operates in a hierarchically organised manner. To obtain a global solution, a neural network is employed to describe the connections of functional modules. The shape and the dimension of each module are simultaneously considered by means of subordinate neural networks, called module nets, to produce the final placement. Either a bitmap net or a closed line net is employed as a module network (see Figure 5.2). Both of these consist of several connected neurons called module elements.

When the bitmap net is used, the relation obtained between net and module is very close. However, this net model requires large computation time. The closed line net is simpler than the bitmap net since each module approximately corresponds to a flexible borderline connected with module elements.

Given a set of modules $M=\{m_1,m_2,...,m_n\}$, each module m_i consists of m module elements O_{ij}: $m_i=\{O_{i1},O_{i2},...,O_{im}\}$. The number of these elements represents the area. The line stands for the shape of the module. After the learning process the net becomes a circular line.

To describe the connection relationship between modules, a high-level mapping neural network, the connection net, is employed. The connectivity between modules is modelled by a weighting system. Using the weighting system, a hierarchical neural network model is obtained (see Figure 5.3).

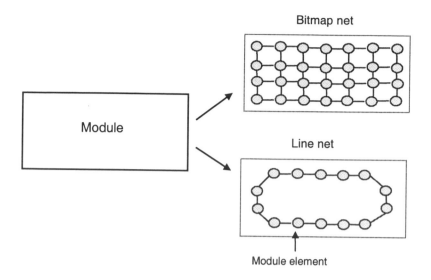

Fig. 5.2 Two neural module nets: bitmap net or line net

To co-ordinate the learning steps, the nets are hierarchically organised with respect to time and topology. Learning consists of two phases. The first phase is for the learning process of the module net (process for all $O_{ij} \in m_i$), and the second phase for the high-level connection net (process for all $m_i \in Z_{me}(s)$). The adaptation of the superior neural network occurs only in the initial phase of learning.

After the learning process has been completed, the positions of module elements are determined in output space A. It is assumed that there is a grid dividing the output space. The minimal distance between the points of the grid and the position of the module element is subsequently computed. After this calculation is carried out for all module elements and all elements are assigned to the grid points a complete placement is obtained. These points generally cluster together so that they represent a module area.

The routing area of the placement is included within the model of each module. The size of the routing area is derived from the number of pins on the module periphery.

To minimise the pin-to-pin total wire length, the rotation and orientation of modules can be performed by using the bitmap nets.

The algorithms were also enhanced in speed and extended to the I/O pad placement and placement on the rectilinear area problems.

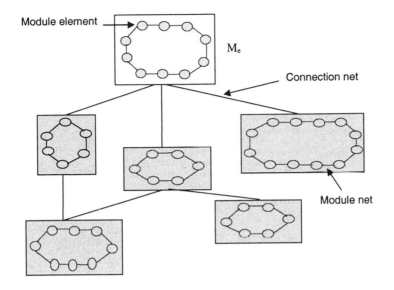

Fig. 5.3 Hierarchical neural placement system

In the pad placement problem, the pads nearest to the pad positions are excited and move toward the positions. The learning rule for pads is:

$$w_i^{(t+1)} = w_i^{(t)} + \gamma L(k) V(i)(x_a - w_i^{(t)}) \quad \forall i \in Z_p(s) \tag{5.12}$$

where w_i denotes the mapping vector for the excited pad p and its connected modules, x_a is the pad position and γ scales the excited value. Z_p is the connection zone of the excited pad p.

Prelocating some large macro cells on the placement carrier causes obstacle regions and the placement area becomes non-rectilinear. In this case, the training inputs are applied frequently in some regions and many modules are activated; but in rarely stimulated regions the modules cannot be activated. The algorithms can be conveniently used for a placement, either on a rectilinear area or in regions with obstacles, when the probability distribution $p(x)$ is restricted to inputs x in the given rectilinear regions.

The learning cycle has two stages. The first stage creates a coarse form of order from a random placement of modules. In this stage, the learning process attempts to cluster the modules according to functional neighbourhoods, i.e., the connectivity between the modules. During this stage the adaptation rate is kept high. After the first stage, the modules within the localised regions of the placement are fine-tuned to the input training vectors. In this stage, much smaller changes of the mapping vectors are made for each module. Therefore the adaptation rate is reduced. To speed up the fine-tuning, a new method is used in this work whereby a threshold is employed to force the modules to move into slots. The technique is effective when the distance between the module position and the slot reaches the threshold d_0:

$$d = (D/n) \leq d_0 \tag{5.13}$$

where d is the average rate of the deviation D and n is the number of modules. At this time most modules reach a close vicinity to their final locations and no long distance movements of cells are necessary.

5.1.3 MHSO2 Experiments

Algorithms were implemented in the C language on a DEC/VAX machine and a SUN workstation and tested for different examples of gate arrays, standard cell and macro cell circuits. Only the results obtained for one problem for each case will be presented here.

To observe clearly whether the result is a global optimum, an example with a strong regular topology was tested. In the optimal placement each module except for those at the periphery is connected to its four neighbours. No diagonal connections exist. For an initial configuration the placement produced after 900 steps is given in Figure 5.4(a). Then, if the placement process continues, an optimal configuration can be reached after 1200 steps (see Figure 5.4(b)). After results were obtained for several experiments it was seen that the convergence speed is affected by the initial placements, but the initial placements do not affect the final results. In the figure L stands for the total wire length.

The MHSO algorithm has an extremely fast convergence for placement with standard cell design style. The placement results of experiments on several different standard cell circuits applying the MHSO algorithm were compared with those obtained with the simulated annealing algorithm. Figure 5.5 is an example with 144 standard cells and 48 pads. The example contains only two-point-nets which have the same connection strength. The placement is performed after 8500 learning steps and takes 28 seconds. The computation time for the same placement by using the simulated annealing is 3.5 minutes.

A number of placement examples were also tested using the MHSO2 algorithm. One of them was obtained using the bitmap net method. The example consists of 15 modules in which 2 modules are fixed previously (IC9 and X1). Figure 5.6 shows the placement result after 100000 learning steps.

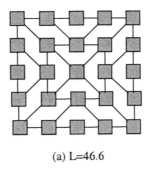

(a) L=46.6

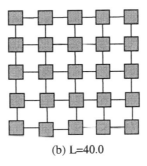

(b) L=40.0

Fig. 5.4 Placement results obtained after (a) 900 steps and (b) 1200 steps

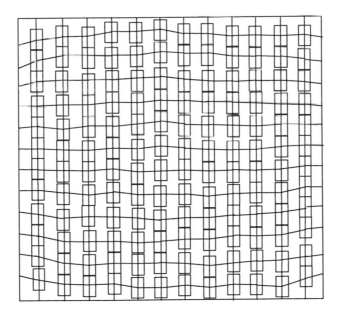

Fig. 5.5 Placement result for a standard cell circuit

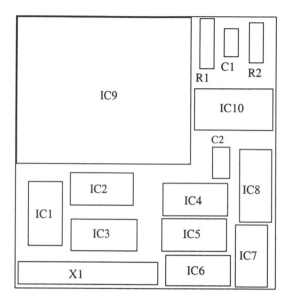

Fig. 5.6 Placement result for an example with 15 modules: placement of modules after 100000 learning steps

5.2 Satellite Broadcast Scheduling using a Hopfield Network

Commercial satellites are usually launched into a geostationary orbit, which means that the satellite maintains its position above the earth's surface. No hand-over is required for stations on the earth. However, the launching cost is high and the satellites require very large antennas. A low-altitude satellite communication system, in contrast, has several advantages. Its satellite power requirement is not high, it has small propagation delays and it enables high-resolution images. The system can also provide global communication coverage including the two polar regions [Funabiki and Nishikawa, 1997].

A low-altitude satellite communication system comprises a set of satellites and a set of ground terminals. The satellites are usually located in a polar orbit with a rather low altitude. The terminals on the ground usually have low-power transmitters and portable antennas. It is necessary to schedule the hand-over of broadcasting operations for the satellites and Bourret *et al.* [1989 and 1990], Ansari *et al.* [1995] and Funabiki and Nishikawa [1997] have applied neural network approaches to solve this problem.

Bourret *et al.* [1989 and 1990] employed a neural network in which neurons are connected in a three-layer model. A sequential search is used to find the optimum schedule and the search is controlled by a competitive activation mechanism based on a dynamic prioritisation of satellites. The sequential search, however, is very time consuming. Also this approach additionally requires a set of distinct priorities of satellites and a set of suitable requests, which are very difficult to determine for large problems.

Ansari *et al.* [1995] used an approach based on the Hopfield neural network to solve the problem. All neurons of the network are completely connected. An energy function is derived for the network and optimisation is achieved by minimising the energy function. Minimisation of the function is carried out by mean field annealing. Since the search for optima is performed in parallel in the global sense, the execution time of the algorithm is short. The approach proposed by Ansari *et al.* does not require the additional information needed in the method by Bourret *et al.* [1989 and 1990]. In this approach, a conventional neuron model is used and some neurons are clamped by an "associative matrix" to reduce the solution space. However, the method necessitates the calculation of an exponent term to update the state of the neurons. This usually requires heavy computation and the time required for convergence increases very rapidly with problem size.

Funabiki and Nishikawa [1997] proposed a neural network approach using simple binary neurons to solve the addressed scheduling problem. This approach reduces the required computation time significantly [Kurokawa and Yamashita, 1994]. In the following sections, the basic definitions related to the problem will be described, the neural network approach proposed by Funabiki and Nishikawa [1997] will be described together with the results.

5.2.1 Problem Definition

Funabiki and Nishikawa [1997] followed the problem definition used by Ansari *et al.* [1995], with some modifications to clearly separate the goal function and the constraints. Communication links between the satellites and ground terminals are provided in a series of time slots. A time slot has a unit time to broadcast information from a satellite to a ground terminal when they are visible to each other. The aim here is to find the broadcasting schedule of satellites with the maximum number of broadcasting time slots under the following set of constraints:

1. A satellite cannot broadcast information to more than one terminal at the same time slot.
2. A terminal cannot receive information from more than one satellite at the same time slot.
3. A satellite cannot be assigned to more time slots than requested from it, unless all the requests are allocated for all the satellites.

4.   A satellite can broadcast only when it is visible from a ground terminal.

A request vector is used to represent the set of requests for the broadcasting time of satellites. The ith element r_i of the vector gives the number of broadcasting time slots requested from satellite i. The solution is called the global solution when all the requests are allocated for all the satellites, and the solution is called a local solution when some requests are not allocated. In the approach, an associative matrix is employed to describe the visibility between the satellites and terminals. The ijk^{th} element d_{ijk} is "1" when satellite i is visible to terminal j at time slot k, and the element d_{ijk} is "0" when it is not visible.

As an example, Figure 5.7(a) presents an associative matrix for a satellite broadcast scheduling problem with three satellites, two terminals, and five time slots. The black squares indicate "1" and the white squares indicate "0". For example, satellite 1 is visible to terminal 1 at time slots 1-4. When request vector (2,3,2) is given to this matrix, the optimum solution is composed of ten broadcasting time slots as shown in Figure 5.7(b).

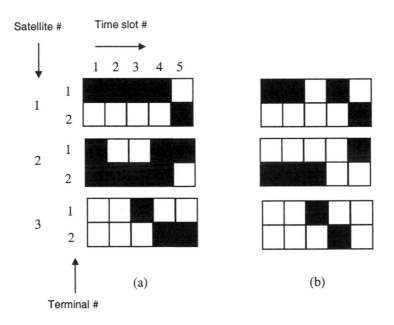

Fig. 5.7 Satellite broadcast scheduling problem: (a) Associative matrix; (b) Optimum solution for the request vector (2,3,2)

5.2.2 Neural-Network Approach

The neural network proposed by Funabiki and Nishikawa [1997] followed the approach by Hopfield and Tank [1985]. However, Funabiki and Nishikawa defined a motion equation instead of the energy function. A three-dimensional network is used that is the same as that described in [Ansari *et al.*, 1995]. The output V_{ijk} of neuron *ijk* represents whether satellite *i* should be assigned to terminal *j* at time slot *k* or not. As in Figure 5.7, a "1" output ($V_{ijk}=1$) represents assignment, and the "0" output ($V_{ijk}=0$) no assignment. Thus, *NxMxL* neurons are required for a problem with *N* satellites, *M* terminals, and *L* time slots. Figure 5.8 shows the neural network composed of 30(3×2×5) neurons, when nearing convergence to the solution corresponding to Figure 5.7(b).

The motion equation prescribes the ability of the neural network to find solutions in a combinatorial optimisation problem. It drives the state of the neural network to a solution. The following motion equation is defined for the neuron *ijk*:

$$\frac{dU_{ijk}}{dt} = -A\sum_{\substack{q=1\\q\neq j}}^{M} V_{iqk} - B\sum_{\substack{p=1\\p\neq i}}^{N} V_{pjk} - Cf(\sum_{q=1}^{M}\sum_{r=1}^{L}V_{iqr} - r_i)$$

$$+ Dh(\sum_{q=1}^{M}\sum_{r=1}^{L}V_{iqr} - r_i) + Eh(\sum_{q=1}^{M}V_{iqk} + \sum_{p=1}^{N}V_{pjk} - 1) \qquad (5.14)$$

where U_{ijk} is the input of the neuron *ijk*, and *A-E* are constant coefficients. To satisfy the fourth constraint, the motion equation is executed only for the visible neuron *ijk* which corresponds to the nonzero element "$d_{ijk}=1$" in the associative matrix.

The *A*-term represents the first of the four constraints in this problem. It discourages neuron *ijk* from having output one if satellite *i* is assigned to a terminal other than *j* at time slot *k*. The *B*-term represents the second constraint. This discourages neuron *ijk* from having output one if terminal *j* is assigned to a satellite other than *i* at time slot *k*. The omega function heuristic is applied on these terms to make the local minimum shallower [Funabiki and Takefuji, 1992a; Funabiki and Takefuji, 1992b; Funabiki and Takefuji, 1993; Funabiki and Nishikawa, 1995]. Instead of the original terms, the slightly modified terms:

$$- A\sum_{\substack{q=1\\q\neq j}}^{M} V_{iqk}V_{ijk} - B\sum_{\substack{p=1\\p\neq i}}^{N} V_{pjk}V_{ijk} \qquad (5.15)$$

are used if (t *mod* 20)<15, where t is the number of iterations.

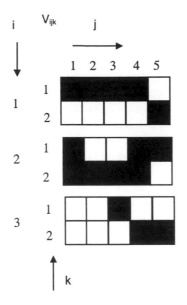

Fig. 5.8 Neural network for the problem in Fig. 5.7

The C-term represents the third constraint. The function $f(x, t)$ is x if $x<0$ and $t<200$, "1" if $x>0$ and $t\geq200$, and "0" otherwise. In the first 200 steps, the C-term encourages at least r_i neurons including neuron ijk for satellite i to have one output for the global solution. If the network does not converge within 200 steps, it will seek a local solution, where the C-term discourages neuron ijk from having output one if more than r_i time slots are assigned to satellite i.

The D-term and E-term are the hill climbing heuristics [Funabiki and Takefuji, 1992a; Funabiki and Takefuji, 1992b; Funabiki and Takefuji, 1993; Funabiki and Nishikawa, 1995]. The function $h(x, t)$ is "1" if $x<0$ and $t<200$, and "0" otherwise. The D-term encourages neuron ijk to have output one if fewer than r_i neurons for satellite i have output one, to reach the global solution. The E-term also encourages neuron ijk to have output one if satellite i can be assigned to terminal j at time slot k without constraint violations in the current state. It aims to increase the number of broadcasting time slots as much as possible.

As the update function of the neuron output, the simple binary McCulloch-Pitts neuron model is adopted [McCulloch and Pitts, 1943]:

if $U_{ijk}>0$ (and $d_{ijk}=1$) then $V_{ijk}=1$ else $V_{ijk}=0$ (5.16)

Note that the update function is also executed only for "visible" neurons to satisfy the fourth constraint.

The iterative computation of the motion equation and the neuron function must be terminated when the state of the neural network satisfies all the constraints and the goal function of the problem. Computation is continued if at least one of the following conditions is not satisfied in any neuron:

1. The A-term is not zero when neuron ijk has output one
2. The B-term is not zero when neuron ijk has output one
3. The C-term is not zero

An additional condition is also checked until 100 iteration steps have elapsed to maximise the number of broadcasting time slots.

4. The E-term is not zero

5.2.3 Simulation Results

In the simulations, Funabiki and Takefuji [1997] used a simulator developed using the synchronous parallel procedure in [Funabiki and Takefuji, 1992a; Funabiki and Takefuji, 1992b; Funabiki and Takefuji, 1993; Funabiki and Nishikawa, 1995], where the states of all neurons are updated simultaneously. A set of parameters including $A=B=C=1$ and $D=E=30$ was empirically selected through simulations based on the selections in [Funabiki and Takefuji, 1992a; Funabiki and Takefuji, 1992b; Funabiki and Takefuji, 1993; Funabiki and Nishikawa, 1995]. In the simulations, they varied the integer values of D and E from 5 to 50 in intervals of 5 to improve the solution quality, while A, B, and C were always fixed at one. The initial input values U were set at uniformly randomised integers between zero and -300, and the initial output values V were set at zero. In the simulations, the seven examples given in Table 5.1 were examined, where 100 simulation runs were performed with different initial input values in each example. Examples 1-3 were the same as in [Ansari *et al.*, 1995], while the other larger examples were randomly generated to evaluate the scaling efficiency of the neural network.

Table 5.2 summarises the simulation results obtained among 100 simulation runs in each of the seven examples. Global solutions were obtained for examples 1, 2, 4, and 5. In example 3, the network found solutions composed of between 52 and 81 broadcasting time slots within 255 iteration steps.

Table 5.1 Specifications of simulated examples

Example	Number of satellites N	Number of terminals M	Number of time slots L	Total number of 1-elements in matrix	Total number of requested time slots	Comment
1	4	3	9	43	8	Example 2 by Ansari
2	4	3	9	44	30	Example 3 by Ansari
3	8	6	18	400	53	Example 4 by Ansari
4	20	16	40	3201	378	New example
5	30	24	60	10755	924	New example
6	40	32	80	25633	1508	New example
7	50	40	100	50042	2394	New example

Table 5.2 Simulation results: (a) Number of iteration steps for convergence; (b) Total number of broadcasting time slots in solutions

Example	Maximum	Minimum	Average
1	129	7	32.6
2	442	278	354.6
3	255	46	105.4
4	288	147	234.2
5	254	127	237.1
6	259	229	246.0
7	249	229	239.0

(a)

Example	Maximum	Minimum	Average
1	18	16	17.4
2	18	17	17.9
3	81	52	78.5
4	615	364	397.7
5	1420	886	918.5
6	1495	1445	1479.7
7	2366	2271	2343.6

(b)

The best solution of 81 broadcasting time slots in example 3 is given in Figure 5.9. In the simulations, for all examples, the network always converged to a near-optimum solution within 500 iterations.

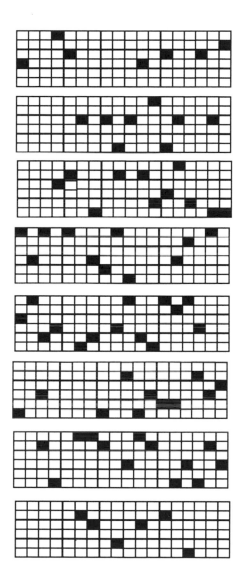

Fig. 5.9 Best solution for the request vector (4, 7, 4, 10, 12, 1, 13, 2)

5.3 Summary

This chapter has presented two examples of engineering applications of neural networks for optimisation. The first example concerns the use of mapping and hierarchical self-organising neural networks for the VLSI placement problem. The second example is the application of the Hopfield neural network to the satellite broadcast scheduling problem.

References

Angeniol, B., De la Croix Vaubois, G. and Le Texier, J.-Y. (1988) Self-organizing feature maps and the traveling salesman problem, *Neural Networks*, Vol.1, pp.289-293.

Ansari, N., Hou, E.S.H. and Yu, Y. (1995) A new method to optimize the satellite broadcasting schedules using the mean field annealing of a Hopfield neural network, *IEEE Trans. Neural Networks*, Vol.6, pp.470-483.

Bourret, P., Goodall, S. and Samuelides, M. (1989) Optimal scheduling competitive activation: application to the satellite antennas scheduling problem, *Proc. Int. Joint Conf. on Neural Networks '89*, pp.1565-1572.

Bourret, P., Rem, F. and Goodall, S. (1990) A special purpose neural network for scheduling satellite broadcasting times, *Proc. Int. Joint Conf. on Neural Networks '90*, pp.1535-1538.

Breuer, M.A. (1977) A class of min-cut placement algorithms, *Proc. 14th ACM/IEEE Design Automation Conf.*, pp.284-290.

Casotto, A. and Sangiovanni-Vincentelli, A. (1987) Placement of standard cells using simulated annealing on the connection machine, *Proc. Int. Conf. Computer-Aided Design*, pp.350-353.

Forbes, R. (1987) Heuristic acceleration of force-directed placement, *Proc. 24th ACM/IEEE Design Automation Conf.*, pp.735-740.

Fort, J.C. (1988) Solving a combinatorial problem via self-organizing process: An application of the Kohonen algorithm to the traveling salesman problem, *Biological Cybernetics*, Vol.59, pp.33-40.

Funabiki, N. and Takefuji, Y. (1992a) A parallel algorithm for channel routing problems, *IEEE Trans. Computer-Aided Design*, Vol.11, pp.464-474.

Funabiki, N. and Takefuji, Y. (1992b) A neural network parallel algorithm for channel assignment problems in cellular radio networks, *IEEE Trans. Vehicle Technology*, Vol.41, pp.430-437.

Funabiki, N. and Takefuji, Y. (1993) A neural network approach to topological via-minimization problems, *IEEE Trans. Computer-Aided Design*, Vol.12, pp.770-779.

Funabiki, N. and Nishikawa, S. (1997) A binary Hopfield neural network approach for satellite broadcast scheduling problems, *IEEE Trans. on Neural Networks*, Vol.8, No.2, pp.441-445.

Funabiki, N. and Nishikawa, S. (1995) An improved neural network for channel assignment problems in cellular mobile communication systems, *IEICE Trans. Communications*, Vol. E78-B, No.8, pp.1187-1196.

Hemani, A. and Postula, A. (1990) Cell placement by self-organization, *Neural Networks*, Vol.3, pp.377-383.

Hopfield, J.J. and Tank, D.W. (1985) Neural computation of decisions in optimization problems, *Biological Cybernetics*, Vol.52, pp.141-152.

Kling, R.M. and Banerjee, P. (1989) ESP: Placement by simulated evolution, *IEEE Trans. Computer-Aided Design*, Vol.CAD-8, pp.245-256.

Kohonen, T. (1989) Speech recognition based on topology preserving neural maps, *Neural Computing Architectures*, Springer-Verlag, Berlin.

Kohonen, T. (1995) *Self-Organizing Map*, Springer-Verlag, Berlin.

Kurokawa, T. and Yamashita, H. (1994) Bus connected neural-network hardware system, *Electronics Letters*, Vol.30, No.12, pp.979-980.

McCulloch, W.S. and Pitts, W.H. (1943) A logical calculus of ideas immanent in nervous activity, *Bull. Math. Biophys.*, Vol.5, p.115.

Qinn, N.R. (1975) The placement problem as viewed from the physics of classical mechanics, *Proc. 12th ACM/IEEE Design Automation Conf.*, pp.173-178.

Rao, D.S. and Patnaik, L.M. (1992) Circuit layout through an analogy with neural networks, *Computer-Aided Design*, Vol.24, pp.251-257.

Sait, S.M. and Youssef, H. (1995) *VLSI Physical Design Automation - Theory and Practice*, IEEE, Piscataway, NJ.

Sechen, C. and Sangiovanni-Vincentelli, A. (1985) The timber wolf placement and routing package, *IEEE J. Solid-State Circuits*, Vol.SSC-20, pp.432-439.

Shahookar, K. and Mazunder, P. (1990), A genetic approach to standard cell placement using meta-genetic parameter optimization, *IEEE Trans. Computer-Aided Design*, Vol.9, pp.500-511.

Yu, M.L. (1989) A study of applicability of Hopfield decision neural nets to VLSI CAD , *Proc. 26th ACM/IEEE Design Automation Conf.*, pp.412-417.

Zhang, C. X. and Mlynski, D.A. (1997) Mapping and Hierarchical self-organizing neural networks for VLSI placement, *IEEE Trans. on Neural Networks*, Vol.8, No.2, pp.299-314.

Appendix 1

Classical Optimisation

This appendix defines some of the terms used in the area of optimisation and presents three classical optimisation algorithms.

A1.1 Basic Definitions

Any problem in which parameter values must be determined, given certain constraints, can be treated as an optimisation problem. The first step in optimisation is to identify a set of parameters, also called decision parameters, a cost function to be minimised and the problem constraints. The cost function gives a lower cost for parameter values that represent a better solution. Restrictions on a solution are called constraints. The constraints of a solution show the values parameters cannot take. Constraints must be expressed in terms of the decision parameters. Some constraints are represented as inequalities and some as equalities. The flowchart of a general design optimisation process is given in Figure A1.1 [Arora, 1989].

The cost function and the constraints are mathematically expressed as follows. Find a vector $x - (x_1, x_2, ... x_n)$ with n components, where x_i represents the value of parameter i:

$$f(x) = f(x_1, x_2, ... x_n) \tag{A1.1}$$

subject to p equality constraints:

$$h_j(x) = h_j(x_1, x_2, ... x_n) = 0 \quad 1 \leq j \leq p \tag{A1.2}$$

and m inequality constraints:

$$g_i(x) = g_i(x_1, x_2, ... x_n) <= 0 \quad 1 \leq i \leq m \tag{A1.3}$$

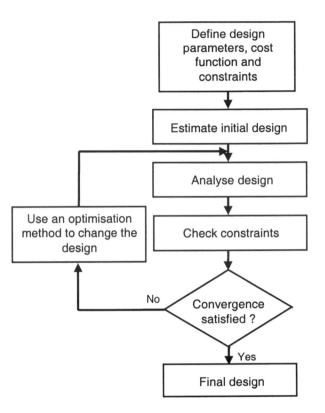

Fig. A1.1 Design optimisation process

In some situations, there may be two or more cost functions. This is called a multi-objective optimisation problem.

The *feasible region* is the set of all solutions to the problem satisfying all the constraints. The *optimal solution* for a minimisation problem is the solution with the smallest cost value in the feasible region. Similarly, for maximisation problems, it is the solution with the largest objective function value. Function $f(x)$ has a global minimum at x^* if:

$$f(x^*) <= f(x) \qquad\qquad (A1.4)$$

$\forall x$ in the feasible region.

Function $f(x)$ has a *local minimum* at x^* if equation (A1.4) holds for all x in a small neighbourhood N of x^* in the feasible region. Neighbourhood N of the point x^* is mathematically defined as:

$N = \{x \mid x \in S \text{ with } \|x - x^*\| < \delta\}$ (A1.5)

for some small δ. Geometrically, it is a small feasible region containing the point x^*. The global and local minima and maxima are shown in Figure A1.2.

The partial derivative of a function $f(x)$ with respect to x_1 at a given point x^* is defined as $d(x^*)/dx_1$. If all partial derivatives are arranged in the form of a column vector, the vector is called the *gradient vector* and is represented as ∇f or grad f. Geometrically, the gradient vector is normal to the tangent plane at point x^* and points in the direction of the maximum increase in the function.

A *Hessian matrix* is obtained by differentiating the gradient vector again. It consists of second partial derivatives of the function $f(x)$. A Hessian matrix is symmetric and plays an important role in the sufficiency conditions for optimality.

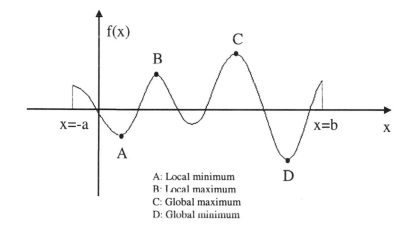

Fig. A1.2 Global and local maximum and minimum points of a multimodal function

A1.2 Classification of Problems

The minimisation or maximisation of $f(x)$ without any constraint on x is called an *unconstrained optimisation*. Conversely, a problem including constraints is called a *constrained optimisation problem*.

If an optimisation problem has linear objective and constraint functions, it is called a *linear programming problem*. An *integer programming problem* is a linear programming problem in which some or all of the variables must be non-negative integers. Otherwise, it is called a *non-integer programming problem.*

The search for an optimal arrangement, grouping, ordering or selection of discrete objects is called *combinatorial optimisation.* A problem having a quadratic objective function and linear constraints is called a *quadratic programming (QP) problem.*

A1.3 Classification of Optimisation Techniques

Optimisation techniques can be broadly classified into two classes: *direct search* and *gradient-based methods.* This classification is based on whether derivative information is used during search. Methods that do not use derivative information are called direct methods whereas gradient-based methods do employ such information. Direct search methods require only the computation of objective function values to select suitable search directions. Since evaluating derivatives of the objective function is usually laborious, and in some cases impossible, this gives an advantage to the direct search class of algorithms. There are three main groups of direct search methods: *tabulation, sequential* and *linear methods* [Arora, 1989].

In tabulation, a set of points is selected using random tabulation or grid tabulation strategies and the objective function is evaluated for each point. The point with the lowest function value is returned as the optimum.

In the sequential method, a geometrical figure that has the same dimension as the argument space is manoeuvred so that it eventually settles around, and then converges onto, a minimum point. Decisions are taken by comparing function values corresponding to the nodes of the figure.

Linear direct search involves sequences of searches along lines in argument space. It consists of two categories of algorithms. The first category works with search directions that are either pre-defined or are determined on the basis of purely locally-derived information (univariate search and pattern move techniques). A simple univariate search offers rapid initial convergence with easily guaranteed stability but suffers from increasingly slow progress as the minimum is approached. The rate of convergence is improved by introducing a pattern move, in which previous and current points are used to infer the direction in which the minimum lies. The second category of algorithms involves a search strategy of a more global nature to exploit the properties of the so-called "conjugate" directions.

Gradient-based methods use the first-order or higher derivatives of the objective function to determine a suitable search direction. There are three main strategies. In the first strategy, first-order derivatives, which represent slope information are used to infer the "downhill" direction of steepest descent. The second strategy uses second-order derivatives, which represent curvature information, to approximate the cost function surface by a quadratic function whose minimum can be easily found (Newton's method). The third method is the use of conjugate, quasi-Newton, techniques [Arora, 1989].

The method of steepest descent is based on the notion of seeking the steepest possible 'downhill' direction. This method uses the negative gradient vector evaluated at the current point. The justification for this is that the positive gradient vector $g(\mathbf{x})$ indicates the direction of steepest ascent of $f(x)$ at point x and its magnitude, denoted by $\|g\|$, gives the slope at x. A new point x^1 is derived from the current point, where $x^1 = x*- \alpha^{(k)}g(x*)$ and the scalar $\alpha^{(k)}$ is chosen to minimise $f(x^1)$ in the direction $-g(x*)$.

The basic steepest descent algorithm is as follows [Arora, 1989]:

Step 1. Estimate the starting solution $\mathbf{x}^*$. Set the iteration number t to zero (t=0) and select a tolerance for the stopping criterion.

Step 2. Calculate the gradient $\mathbf{g}^{(t)}$ at $\mathbf{x}^{(t)}$ and then calculate $\|\mathbf{g}^{(t)}\|$. If $\|\mathbf{g}^{(t)}\|$ is less than the tolerance then stop the search; otherwise continue.

Step 3. Calculate the direction of search in the search space using $\mathbf{d}^{(t)} = -\mathbf{g}^{(t)}$

Step 4. Determine a new point $\mathbf{x}^{(t+1)} = \mathbf{x}^{(t)} + \alpha^{(t)}\mathbf{d}^{(t)}$, where $\alpha^{(t)}$ is the step size calculated to minimise $f(\mathbf{x}^{(t)} + \alpha^{(t)}\mathbf{d}^{(t)})$.

Step 5. Set t = t+1 and go to step 2.

In Newton's method, it is assumed that the first and second derivatives exist. The gradient g and Hessian H are computed at $x*$ and used to predict the position of the minimum of the quadratic approximating surface. The predicted minimum is expressed as $x^1 = x* + \Delta x$, where $\Delta x = -\alpha H(x*)^{-1}g(x*)$. x^1 is a minimum if $\|H\|$ is positive. The basic steps of the modified Newton's method are [Arora, 1989]:

Step 1. Estimate the starting solution $\mathbf{x}^*$. Set iteration number t to zero and select a tolerance for the stopping criterion.

Step 2. Calculate the gradient $\mathbf{g}^{(t)}$ at $\mathbf{x}^{(t)}$ and then calculate $\|\mathbf{g}^{(t)}\|$. If $\|\mathbf{g}^{(t)}\|$ is less than the tolerance then stop the search; otherwise continue.

Step 3. Compute the Hessian $\mathbf{H}(\mathbf{x}^{(t)})$.

Step 4. Calculate the direction of search using $\mathbf{d}^{(t)} = -\mathbf{H}^{-1}\mathbf{g}^{(t)}$

Step 4. Determine a new point $\mathbf{x}^{(t+1)} = \mathbf{x}^{(t)} + \alpha^{(t)}\mathbf{d}^{(t)}$, where $\alpha^{(t)}$ is the step size calculated to minimise $f(\mathbf{x}^{(t)} + \alpha^{(t)}\mathbf{d}^{(t)})$.

Step 6. Set t = t+1 and go to step 2.

Quasi-Newton methods are based on the Newton technique but avoid the explicit evaluation of the Hessian and its inversion. These methods need the computation of only first derivatives. They use the first-order derivatives to generate approximations for the Hessian matrix. The basic idea is to update the current approximation using changes in the solution and in the gradient vector. Thus, H^{-1} is replaced by H^r representing an approximation to H^{-1} after r iterations. The most commonly-used algorithm of this type is the Davidon-Fletcher-Powell (DFP) algorithm [Arora, 1989]. The basic steps of this algorithm are:

Step 1. Estimate the starting solution **x***. Choose a symmetric positive definite matrix A* as an estimate for the inverse of the Hessian matrix of the objective function. Set the iteration number t to zero and select a tolerance for the stopping criterion.

Step 2. Calculate the gradient $\mathbf{g}^{(t)}$ at $\mathbf{x}^{(t)}$ and then calculate $\|\mathbf{g}^{(t)}\|$. If $\|\mathbf{g}^{(t)}\|$ is less than the tolerance then stop the search; otherwise continue.

Step 3. Calculate the direction of search using $\mathbf{d}^{(t)} = -A^{(t)}\mathbf{g}^{(t)}$

Step 4. Determine a new point $\mathbf{x}^{(t+1)} = \mathbf{x}^{(t)} + \alpha^{(t)}\mathbf{d}^{(t)}$, where $\alpha^{(t)}$ is the step size calculated to minimise $f(\mathbf{x}^{(t)} + \alpha^{(t)}\mathbf{d}^{(t)})$.

Step 5. Update the matrix $A^{(t)}$. $A^{(t+1)} = A^{(t)} + B^{(t)} + C^{(t)}$; where B and C are called correction matrices, $B^{(t)} = \mathbf{s}^{(t)}\mathbf{s}^{(t)T}/(\mathbf{s}^{(t)}.\mathbf{y}^{(t)})$, $C^{(t)} = -\mathbf{z}^{(t)}\mathbf{z}^{(t)T}/(\mathbf{y}^{(t)}.\mathbf{z}^{(t)})$ where $\mathbf{s}^{(t)} = \alpha^{(t)}\mathbf{g}^{(t)}$ (change in solution), $\mathbf{y}^{(t)} = \mathbf{g}^{(t+1)} - \mathbf{g}^{(t)}$ (change in gradient) and $\mathbf{z}^{(t)} = A^{(t)}\mathbf{y}^{(t)}$

Step 6. Set t = t+1 and go to step 2.

From the above algorithms, it can be seen that an optimisation process can be broken into two phases: search direction $(d^{(t)})$ and step length $(\alpha^{(t)})$ determination. After the search direction is determined, step size determination is a one-dimensional minimisation problem. Three simple algorithms to solve the step size problem are *equal interval search*, *golden sectional search* and *polynomial interpolation*. The basic idea of these techniques is to reduce successively the interval of uncertainty to a small acceptable value.

As with unconstrained optimisation problems, several methods have been investigated for constrained optimisation. The most well-known is the simplex method used to solve linear programming problems. This method is an extension of the standard Gauss-Jordan elimination process for solving a set of linear equations. In some situations, there may be multiple objectives and no feasible solution satisfying all objectives. In this case, goal programming techniques are used.

Most integer programming problems are solved by using the branch-and-bound technique. Branch-and-bound methods find the optimal solution to an integer programming problem by efficiently enumerating the points in a feasible region.

In most constrained optimisation methods, Taylor series expansion is employed to linearise non-linear problems. The linearised subproblem is transformed to a QP problem, having a quadratic cost function and linear constraints. QP problems can be solved using an extension of the simplex method. Two well-known methods for constrained problems are constrained steepest descent and quasi-Newton methods, which are the modified versions of the methods described above [Arora, 1989].

References

Arora, J.S. (1989) *Introduction to Optimum Design*, McGraw-Hill, New York.

Appendix 2

Fuzzy Logic Control

This appendix gives the basic concepts of fuzzy sets and fuzzy logic control (FLC).

A2.1 Fuzzy Sets

Fuzzy logic control is based on fuzzy set theory [Zadeh, 1965]. This section introduces fuzzy set theory and some basic fuzzy set operations. Approximate reasoning using fuzzy relations is explained [Zadeh, 1973]. Finally, FLC is briefly described and some of its engineering applications summarised.

A2.1.1 Fuzzy Set Theory

Fuzzy set theory may be considered an extension of classical set theory. While classical set theory is about "crisp" sets with strict boundaries, fuzzy set theory is concerned with "fuzzy" sets whose boundaries are not defined sharply.

In classical set theory, an element (u) either belongs to a set A or does not, i.e. the degree to which element u belongs to set A is either 1 or 0. However, in fuzzy set theory, the degree of belonging of an element u to a fuzzy set $\underset{\sim}{A}$ is a real number between 0 and 1. This is denoted by $\mu_A(u)$, the grade of membership of u in $\underset{\sim}{A}$. Fuzzy set $\underset{\sim}{A}$ is a fuzzy set in U, the "universe of discourse" or "universe", which includes all objects under consideration. $\mu_A(u)$ is 1 when u is definitely a member of $\underset{\sim}{A}$ and $\mu_A(u)$ is 0 when u is definitely not a member of $\underset{\sim}{A}$. Figures A2.1(a) and (b) show a crisp set A and fuzzy set $\underset{\sim}{A}$ of numbers near to 2.

Definition 1. Fuzzy set $\underset{\sim}{A}$ in U is a set of ordered pairs:

$$\underset{\sim}{A} = \{ (u, \mu_A(u)) \mid u \in U \} \tag{A2.1}$$

Definition 2. The support of a fuzzy set $\underset{\sim}{A}$, $S(\underset{\sim}{A})$, is the crisp set of all $u \in U$ such that $\mu_A(u) > 0$.

Definition 3. The set of elements that belong to fuzzy set $\underset{\sim}{A}$ to at least a given degree α is called the α-level set.

$$\underset{\sim}{A}_\alpha = \{ u \in U \mid \mu_A(u) \geq \alpha \} \tag{A2.2}$$

$\underset{\sim}{A}_\alpha$ is called a "strong α-level set" or "strong α cut" if $\mu_A(u) \geq \alpha$.

Definition 4. A fuzzy set $\underset{\sim}{A}$ defined in U is called a fuzzy singleton if the support of $\underset{\sim}{A}$ contains only one element and its membership value is 1.

Definition 5. If $\underset{\sim}{A}$ and $\underset{\sim}{B}$ are fuzzy sets in universes U and V, respectively, the Cartesian product of $\underset{\sim}{A}$ and $\underset{\sim}{B}$ is a fuzzy set in the product space UxV with membership function:

$$\mu_{A \times B}(u, v) = \text{Min} \{ \mu_A(u), \mu_B(v) \} \text{ where } (u, v) \in (UxV) \tag{A2.3}$$

A2.1.2 Basic Operations on Fuzzy Sets

Fuzzy sets are defined by membership functions. Therefore, their set operations are also usually defined in terms of membership functions. The most commonly-adopted definitions are those used by Zadeh [1965]. They are given below for the intersection, union and complement operations.

Definition 6. The intersection $\underset{\sim}{C}$ of fuzzy sets $\underset{\sim}{A}$ and $\underset{\sim}{B}$ is defined by:

$$\mu_C(u) = \text{Min} \{ \mu_A(u), \mu_B(u) \} \text{ where } u \in U \tag{A2.4}$$

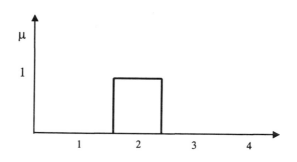

Fig. A2.1 (a) Crisp set of numbers "near" to 2

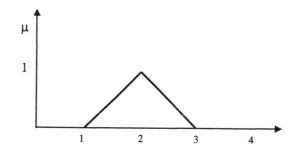

Fig. A2.1 (b) Fuzzy set of numbers "near" to 2

Definition 7. The union D of fuzzy sets A and B is given by:

$$\mu_D (u) = Max \{ \mu_A (u), \mu_B (u) \} \text{ where } u \in U \qquad (A2.5)$$

Definition 8. The complement $\bar{A}$ of fuzzy set A is defined by:

$$\mu_{\bar{A}} (u) = 1 - \mu_A (u) \text{ where } u \in U \qquad (A2.6)$$

Let A and B be the fuzzy sets of numbers near to 2 and near to 3 (Figure A2.2(a)). The intersection and the union of these two sets are given in Figures A2.2(b) and (c), respectively. Figure A2.2(d) shows the complement of fuzzy set A.

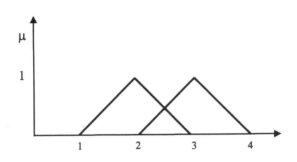

Fig. A2.2 (a) Two fuzzy sets

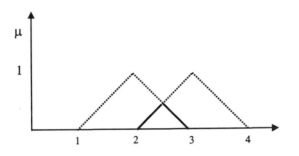

Fig. A2.2 (b) Intersection of the two fuzzy sets in Figure 2.2(a)

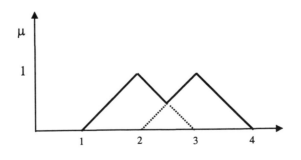

Fig. A2.2 (c) Union of the fuzzy sets in Figure 2.2(a)

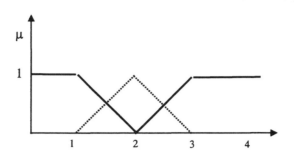

Fig. A2.2 (d) Complement of the fuzzy set representing numbers "near" to 2

A2.2 Fuzzy Relations

A fuzzy system consists of a set of rules called "fuzzy rules". A simple fuzzy rule has the following form:

$$\textit{"IF } A_i \textit{ THEN } B_i\text{"}$$

where the antecedent A_i and the consequent B_i are qualitative expressions.

An example of a fuzzy rule for a fuzzy logic controller is:

IF the error between the desired and actual outputs of the process is Positive_Big

THEN the change in the controller output (the input of the process) is Negative_Big

Here, A_i represents "error is Positive_Big" and B_i is that the "change in controller output is Negative_Big". Such a rule expresses a conditional relation R_i between A_i and B_i. R_i is generally identified with the previously defined Cartesian product, i.e.:

$$R_i = A_i \times B_i = \sum_{U \times V} \mu_{R_i}(u, v) / (u, v) \tag{A2.7}$$

R_i can be expressed as a two dimensional matrix with elements $\mu_{R_i}(u,v)$:

$$
\mu_{R_i}(u,v) = \begin{bmatrix} \mu_{R_i}(u_1, v_1) & \cdots & \mu_{R_i}(u_1, v_q) \\ \\ \mu_{R_i}(u_p, v_1) & \cdots & \mu_{R_i}(u_p, v_q) \end{bmatrix} \tag{A2.8}
$$

If n fuzzy rules can be expressed as relations R_1 to R_n, an overall fuzzy relation R representing all rules can be constructed by combining these individual relations. By employing the union operation, R is defined as:

$$
R = \bigcup_{i=1}^{n} R_i \tag{A2.9}
$$

A2.3 Compositional Rule of Inference

This rule enables a fuzzy system to infer an output corresponding to an arbitrary input not explicitly covered by the individual fuzzy rules using the separate relation matrices R_1 to R_n or the overall relation matrix R. Therefore, this rule describes a method of approximate reasoning [Zadeh, 1975].

In its simplest form, for one input variable and one output variable, the rule is expressed as:

$$
b = a \circ R \tag{A2.10}
$$

Here, a is an arbitrary input and is not necessarily the antecedent of any of the relations R_1 to R_n. b is the inferred process output, given as:

$$
b = \sup(a \times R) \tag{A2.11}
$$

The compositional rule of inference can be regarded as a generalised "modus ponens" rule [Lee, 1990]. This rule permits the inference that the output of a process is b', based on the relation "**If** x *is* a **then** y *is* b " and the fact "x *is* a'". In the case of the compositional rule of inference, the variables are fuzzy and in the case of the conventional modus ponens rule, the variables are crisp.

The compositional rule of inference has two advantages. First, it enables approximate and uncertain inputs to be dealt with. Second, it allows the number of relations required to describe the behaviour of a process to be kept small because it is not necessary to have rules specifically for all possible inputs.

A2.4 Basic Structure of a Fuzzy Logic Controller

Fuzzy set theory was first applied to the design of a controller for a dynamic plant by Mamdani [1974]. Figure A2.3 shows the basic structure of a FLC. It consists of four principal units: fuzzification, knowledge base, decision-making (computation) and defuzzification units. Since data manipulation in a FLC is based on fuzzy set theory, a fuzzification process is required to convert the measured "crisp" inputs to "fuzzy" values. First, the fuzzification unit maps the measured values of input variables into corresponding universes of discourse. Second, it converts the mapped input data into fuzzy sets based on fuzzy values, such as Positive_Big (PB), Negative_Small (NS), etc.

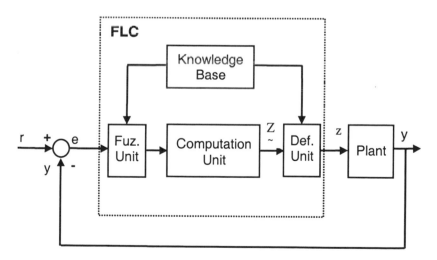

Fig. A2.3 Basic structure of an FLC using only input-output error

The knowledge base contains a set of rules or a relation matrix representing those rules and the information regarding the normalisation and/or discretisation of universes of discourse, fuzzy partitions of input and output spaces, and membership functions defining fuzzy values. Definitions used for the manipulation of fuzzy data are also stored in this unit.

The relation between the input and output of a simple fuzzy logic controller is represented as a set of fuzzy rules or a corresponding overall fuzzy relation matrix. In other words, the rules, or relation matrices, are used to control the process. The main problem in FLC design is how to obtain these fuzzy rules. Various methods have been developed for this purpose. These methods can be classified into four main groups:

1. Eliciting Rules From Operators. This is a common method for constructing the rule base of FLCs. However, the method has some problems. First, a skilled operator may not be available. Second, two operators may not describe their experience correctly even if they have a good knowledge of the process. Researchers who have employed this method include [Kickert and Van Nauta Lemke, 1976; Umbers and King, 1980; Haruki and Kikuchi, 1992; Sasaki and Akiyama, 1988].

2. Observing Operators' Control Actions. Where the skills of operators are important, their actions are observed and described in the form of fuzzy rules [Takagi and Sugeno, 1983]. This method is difficult and time consuming to implement.

3. Using a Fuzzy Model of the Process. The behaviour of the process to be controlled is expressed in the form of fuzzy rules. After modelling the process, the rules for the controller are obtained by searching the controller output to determine the corresponding input that will bring the present state of the process to some desired state [Czogala and Pedrycz, 1982]. This method is indirect and prone to inaccuracies.

4. Using Learning. The knowledge bases designed using the above three methods are static and so are not able to adapt to changes. A learning system can create its own rules to control a process and modify them according to experience [Procyk and Mamdani, 1979; Shao, 1988; Yamazaki, 1982; Moore and Harris, 1992; Berenji, 1992; Kosko, 1992]. Generally, such learning systems are complex to implement.

The decision-making unit simulates the inference mechanism in humans. It produces fuzzy control actions using fuzzy implication. Fuzzy implication is the application of a fuzzy relation and the rule of inference. Researchers have experimented with different methods for implication and inference to determine the most suitable methods for fuzzy control problems [Yamazaki, 1982; Lee,

1990b]. For example, using eight intuitive criteria, Lee tested seven widely-employed implication and inference methods and concluded that the following two are the best for FLCs.

1. Max-Min Rule. This rule is also called Mamdani's minimum operation rule since it was first applied by Mamdani [1974]. Using this rule, for implication and inference, the following can be written:

$$\mu_R(u,v) = Min \ (\mu_A(u), \ \mu_B(v)) \tag{A2.12}$$

$$\mu_{B'}(u,v) = \underset{U}{Max} \ \{ \ Min \ (\mu_{A'}(u), \ \mu_R(u,v)) \ \} \tag{A2.13}$$

2. Max-Product Rule. This rule is obtained by replacing the Min operation in the previous rule with the product operation. It was proposed by Larsen [1980]:

$$\mu_R(u,v) = (\mu_A(u) \cdot \mu_B(v)) \tag{A2.14}$$

$$\mu_{B'}(u,v) = \underset{U}{Max} \ \{ \ (\mu_{A'}(u) \cdot \mu_R(u,v)) \ \} \tag{A2.15}$$

As before, in the above four equations, $u \in U$, $v \in V$ and $(u, v) \in U \times V$, and A and B are the fuzzy values of the input and output variables in a control rule. A' is any fuzzy input to the controller and B' is the corresponding fuzzy output.

The output of the decision-making unit is a fuzzy set. However, a deterministic value is generally required as the input to the process. That is, an interface unit between the process and the decision-making unit is necessary. Several procedures have been proposed for this defuzzification task. The following two are commonly used in control applications.

1. Average of Maxima Criterion. The deterministic output z is calculated by taking the weighted average of elements v_i having the locally highest membership values, $\mu_H(v_i)$:

$$z = (\sum_{i=1}^{n} \mu_H(v_i) \ (v_i)) / (\sum_{i=1}^{n} \mu_H(v_i)) \tag{A2.16}$$

where v_i is the support value at which the membership function reaches a local maximum value μ_H, and n is the number of such support values.

2. Centre of Area Method. This is a popular defuzzification method. Generally, it gives a better steady-state performance than other methods [Yamazaki, 1982]. The crisp output is obtained using the following formula:

$$z = (\sum_{i=1}^{n} \mu(v_i)(v_i)) / (\sum_{i=1}^{n} \mu(v_i)) \qquad (A2.17)$$

where n is the number of support values of the fuzzy set, v_i is a support value and μ is the membership degree of v_i.

A2.5 Studies in Fuzzy Logic Control

Studies in fuzzy logic control can be classified into three main groups:

1. Theory of Fuzzy Logic Control. These studies deal with stability and controllability analysis of fuzzy logic controllers and mathematically-based comparisons between fuzzy logic controllers and other classical or modern controllers. This group also includes studies of the effect of the inference mechanism, defuzzification, fuzzification, rules and membership functions on the performance of the controller [Abdelnour and Cheung, 1992; Braae and Rutherford, 1977; Bouslama and Ichikawa, 1992; Mizumoto and Zimmermann, 1982; Pedrycz, 1990; Singh, 1992; Ying *et al.*, 1990; Cao *et al.*, 1990].

2. Design Methods for Fuzzy Logic Controllers. These studies are concerned with three areas. First, finding new ways to obtain the membership functions, rules or relation matrices. Second, developing new efficient defuzzification procedures and inference mechanisms. Third, designing multi-input-output fuzzy controllers [Czogala and Pedrycz, 1982; Peng, 1990; Chen, 1989; Gupta *et al.*, 1986; Takagi and Sugeno, 1983; Dote, 1990; Gupta and Qi, 1991; Kosko, 1992; Pham and Karaboga, 1991a; Moore and Harris, 1992; Berenji, 1992; Karaboga, 1996].

3. Practical Aspects of Fuzzy Logic Control. These studies include hardware and software design studies, implementations for different kinds of industrial systems and practical comparisons of classical and modern control techniques in real-time applications [Chen *et al.*, 1992; Eichfeld *et al.*, 1992; Hirota and Ozawa, 1989; Ishizuka *et al.*, 1992; Lim and Takefuji, 1990; Yamakawa, 1988; Yamakawa and Kabuo, 1988].

References

Abdelnour, G. and Cheung, J.Y. (1992) Transient response analysis of a fuzzy logic controller, *Proc. IEEE Int. Conf. on Fuzzy Systems*, San Diego, CA, pp.511-518.

Berenji, H. R. (1992) A reinforcement learning-based architecture for fuzzy logic control, *Int. J. of Approximate Reasoning*, Vol.6, No.2, pp.267-292.

Bouslama, F. and Ichikawa, A. (1992) Fuzzy control rules and their natural control laws, *Fuzzy Sets and Systems*, Vol.48, pp.65-86.

Braac, M. and Rutherford, D.A. (1977) Theoretical and linguistic aspects of the fuzzy logic controller, *Automatica*, Vol.15, pp.553-577.

Cao, Z., Kandel, A., Li, L. and Han, J. (1990) Mechanism of fuzzy logic controller, *Proc. 1st Int. Conf. Uncertainty Modelling and Analysis*, pp.603-607.

Chen, Y.Y. (1989) Rules extraction for fuzzy control systems, *Proc. IEEE Int. Conf. on Systems, Man and Cybernetics*, Vol.2, pp.526-527.

Chen, J.J., Chen, C.C. and Tsao, H. W. (1992) Tunable membership function circuit for fuzzy control systems using CMOS technology, *Electronics Letters*, Vol.28, No.22, pp. 2101-2103.

Czogala, E. and Pedrycz, W. (1982) Fuzzy rules generation for fuzzy control, *Cybernetics and Systems*, Vol.13, pp.275-293.

Dote, Y. (1990) Fuzzy and neural network controller, *16th Annual Conf. IEEE Industrial Electronics Society Pacific*, Grove, CA, Vol.2, pp.1314-1343.

Eichfeld, H., Lohner, M. and Muller, M. (1992) Architecture of a CMOS fuzzy logic controller with optimized memory organization and operator design, *Proc. IEEE Int. Conf. Fuzzy Systems*, San Diego, CA, pp.1717-1723.

Gupta, M.M., Kiszka, J.B. and Trojan, G.M. (1986) Multi variable structure of fuzzy control systems, *IEEE Trans. on Systems, Man and Cybernetics*, Vol.SMC-16, No.5, pp.638-656.

Gupta, M.M and Qi, J. (1991) Fusion of fuzzy logic and neural networks with applications to decision and control problems, *Proc. American Control Conference*, Vol.1, pp.30-31.

Haruki, T. and Kikuchi, K. (1992) Video camera system using fuzzy logic, *IEEE Trans. on Consumer Electronics*, Vol.38, No.3, pp.624-634.

Hirota, K. and Ozawa, K. (1989) Fuzzy flip-flop and fuzzy registers, *Fuzzy Sets and Systems* Vol.32, pp.139-148.

Ishizuka, O., Tanno, K., Tang, Z. and Matsumoto, H. (1992) Design of a fuzzy controller with normalization circuits, *Proc. IEEE Conf. on Fuzzy Systems*, San Diego, CA, pp.1303-1308.

Karaboga, D. (1996) Design of fuzzy logic controllers using tabu search algorithm, *Biennial Conference of the North American Fuzzy Information Processing Society (NAFIPS'96)*, University of California, Berkeley, CA, pp.489-491.

Kickert, W.J.M. and Van Nauta Lemke, H.R. (1976) Application of a fuzzy controller in a warm water plant, *Automatica* Vol.12, No.4, pp.301-308.

Kosko, B. (1992) *Neural Networks and Fuzzy Systems*, Prentice-Hall, Englewood Cliffs, NJ.

Larsen, P.M. (1980) Industrial application of fuzzy logic control, *J. Man Machine Studies*, Vol.12, No.1, pp.3-10.

Lee, C.C. (1990) Fuzzy logic in control systems: Fuzzy logic controller, Part I, *IEEE Trans. on Systems, Man and Cybernetics*, Vol.20, No.2, pp.404-418.

Lim, M.H. and Takefuji, Y. (1990) Implementing fuzzy rule-based systems on silicon chips, *IEEE Expert*, February, pp.31-45.

Mamdani, E.H. (1974) Applications of fuzzy algorithms for control of simple dynamic plant, *Proc. IEE*, Vol.121, No.12, pp.1585-1588.

Mizumoto, M. and Zimmermann, H.-J. (1982) Comparison of fuzzy reasoning methods, *Fuzzy Sets and Systems*, Vol.8, pp.253-283.

Moore, C.G. and Harris, C.J. (1992) Indirect adaptive fuzzy control, *Int. J. Control*, Vol.56, No.2, pp.441-468.

Pedrycz, W. (1990) Fuzzy systems: Analysis and synthesis from theory to applications, *Int. J. General Systems*, Vol.17, pp.139-156.

Peng, X.T. (1990) Generating rules for fuzzy logic controllers by functions, *Fuzzy Sets and Systems*, Vol.36, pp.83-89.

Pham, D.T. and Karaboga, D. (1991) Optimum design of fuzzy logic controllers using genetic algorithms, *J. of Systems Engineering*, Vol.1, No.2, pp.114-118, 1991.

Procyk, T.J. and Mamdani, E.H. (1979) A linguistic self-organizing process controller, *Automatica*, Vol.15, pp.15-30.

Sasaki, T. and Akiyama, T. (1988) Traffic control process of expressway by fuzzy logic, *Fuzzy Sets and Systems,* Vol.26, pp.165-178.

Shao, S. (1988) Fuzzy self-organizing controller and its application for dynamic process, *Fuzzy Sets and Systems,* Vol.26, pp.151-164.

Singh, S. (1992) Stability analysis of discrete fuzzy control systems, *Proc. IEEE Int. Conf. on Fuzzy Systems*, San Diego, CA, pp.527-534.

Takagi, T and Sugeno, M. (1983) Derivation of fuzzy control rules from human operator's control actions, *Proc. Fuzzy Information, Knowledge Representation and Decision Analysis Symp, IFAC*, Marseille, pp.55-60.

Umbers, I.G. and King, P.J. (1980) An analysis of human-decision making in cement kiln control and the implications for automation, *Int. J. Man Machine Studies*, Vol.12, No.1, pp.11-23.

Yamakawa, T. (1988) High-speed fuzzy controller hardware system: The mega-FIPS machine, *Information Sciences*, Vol.45, pp.113-128.

Yamakawa, T. and Kabuo, H. (1988) A programmable fuzzifier integrated circuit-synthesis, design and fabrication, *Information Sciences*, Vol.45, pp.75-112.

Yamazaki, T. (1982) *An Improved Algorithm for a Self Organising Controller*, Ph.D. Thesis, Queen Mary College, University of London, U.K.

Ying, H., Siler, W. and Buckley, J.J. (1990) Fuzzy control theory: A nonlinear case, *Automatica*, Vol.26, No.3, pp.513-520.

Zadeh, L.A. (1965) Fuzzy sets, *Information Control*, Vol.8, pp.338-353.

Zadeh, L. A. (1975) The concept of a linguistic variable and its application to approximate reasoning-III, *Information Sciences*, Vol.9, pp.43-80.

Zimmermann, H.-J. (1991) *Fuzzy Set Theory and Its Applications* (2nd ed.) Kluwer, Dardrecht-Boston.

Appendix 3

Genetic Algorithm Program

```
/* The standard genetic algorithm defined by Grefenstette*/

#include <stdio.h>
#include <stdlib.h>
#include <string.h>
#include <time.h>
#include <math.h>
#include <dos.h>

#define range_rand(min, max) ((rand() % (int)(((max)+1) - (min))) + (min))
#define unit_rand() (double)(rand())/RAND_MAX
#define maxbits 24
#define maxindiv 50      /*population size */
#define NumOfTrials 1
#define maxgen 200      /* maximum number of generations */
#define p_cross 0.85      /* crossover rate (c) */
#define p_mutation 0.015      /*mutation rate (m) */
#define maxpossible 1.0      /*upper limit of the fitness value */
#define MAXIND 400
#define MAXWORDS 4
#define gengap 1.0      /* generation gap (g) */
#define window 1   /* scaling window (w) */
#define strategy 1   /* selection: 1 = elitist, 0 = pure. */
#define two32 4294967296.0
#define two31 2147483648.0
#define VALUELIM 200

char FileStep[2];
char FileName[12];
char *FileName1="gen2_";


int best, last, gencount, gencut1, gencut2, FileNum;
int maskcount, words;
long crossmask, invmask, inputmask;
```

```
long bitselect[32], bitmask[32], upper[32], lower[32];
double gen_scale, genstrength;
unsigned long pool[MAXWORDS][MAXIND],
source[MAXWORDS][MAXIND],EvalStr;
double value[MAXIND], partsum[MAXIND], rating[MAXIND];
double fmin[6], fmin1 ;
double result, x[2];
double fit,f;
int kt,i,j,k,m,n;

double evalfunc(unsigned long EvalString)
{
      double evalRes;
      unsigned int Resolution;

      Resolution=(pow(2,12)-1);
      x[0] = EvalString & Resolution;
      x[1] = (EvalString >> 12);
      x[0] = 2.047*((x[0]-Resolution/2)*2/Resolution);
      x[1] = 2.047*((x[1]-Resolution/2)*2/Resolution);
      evalRes=10000.0 - (100.0*pow(pow(x[0],2.0)
            -x[1],2.0)+pow(1.0- x[0],2.0));
      return(evalRes);
}


void convert()
{
      f=0.0;
      for (kt = 0; kt < maxindiv; kt++)
      {
            EvalStr = pool[0][kt] ;
            value[kt]=evalfunc(EvalStr);
      }
}


/* this routine simulates the spinning of a roulette wheel in order to
* select the seeded members of the next generation. the width of each
* individual's slot on the wheel depends on its fitness compared to the
* rest of its generation. The search performed to find the correct
* source index (mid_index) is a binary search. */
void seeded_selection()
{
      int high, low, mid, i, tc;
      int found;
      double spin;
```

```
for (tc = gencut2 - 1; tc < maxindiv; tc++)
{
    spin = unit_rand() * partsum[maxindiv - 1];
    high = maxindiv - 1;
    low = 0;
    found = 0;
    mid = (high + low)/2;
    while ((found == 0) && (high >= low) && (mid > 0))
    {
        if ((partsum[mid] > spin) && (partsum[mid - 1] <= spin))
            found = 1;
        else
            if (partsum[mid] > spin)
                high = mid - 1;
            else low = mid + 1;
                mid = (high + low)/2;
    }
    for (i = 0; i < words; i++)
        pool[i][tc] = source[i][mid];
}
}

/* This routine uses the generation gap (g) to select at random
 * a portion of an existing population to survive intact to the
 * next generation. */
void random_selection()
{
    int choice, randcount, i;
    while (gencut1 > 0)
    {
        for (randcount = 0; randcount < gencut1; randcount++)
        {
            do
                choice = range_rand(0, maxindiv-1);
            while (rating[choice] < 0.0);
            for (i = 0; i < words; i++)
                pool[i][randcount] = source[i][choice];
        }
    }
}

/* This procedure performs the crossover operation on a generation. The
 * crossover probability is used to decide if each individual is to be
 * affected. If so, then a mate is selected and crossover takes place.
 * Whether to use the head or tail of the first parent is decided by the
 * tossing of a coin. */
```

```
void crossover()
{
    int cc, crosspoint, crossword, i, nominee;
    long a, b;
    for (cc = 0; cc < maxindiv; cc++)
    {
        if (unit_rand() <= p_cross)
        {
            do
                nominee = range_rand(0, maxindiv-1);
            while (nominee == cc);
            crosspoint = range_rand(0, maxbits - 2) + 1;
            crossword = crosspoint/32;
            crosspoint %= 32;
            if (crosspoint == 0)
                crosspoint = 32;
            else crossword += 1;
            if (unit_rand() <= 0.5)
            {
                a = upper[crosspoint-1] & pool[crossword-1][cc];
                b = lower[crosspoint-1] & pool[crossword-1][nominee];
                pool[crossword-1][cc] = a | b;
                for (i = 0; i < crossword-1; i++)
                    pool[i][cc] = pool[i][nominee];
            }
            else
            {
                a = upper[crosspoint-1] & pool[crossword-1][nominee];
                b = lower[crosspoint-1] & pool[crossword-1][cc];
                pool[crossword-1][cc] = a | b;
                for (i = crossword; i < words; i++)
                    pool[i][cc] = pool[i][nominee];
            }
        }
    }
}

/* This procedure performs the mutation operation on each
 * member of a new generation. Each bit is mutated with the
 * probability of p_mutation. */

void mutation()
{
    int i, j, k;
    long mutmask;
    for (i = 0; i < maxindiv; i++)
```

```
{
          for (k = 0; k < words; k++)
          {
            mutmask = 0;
              for (j = 0; j < 32; j++)
              {
                    mutmask <<= 1;
                    if (unit_rand() <= p_mutation)
                        mutmask |= 1;
              }
              if ((k == words-1) && (last>0))
                    mutmask &= lower[last - 1];
              pool[k][i] ^= mutmask;
          }
      }
}

/* This procedure calculates the fitness rating of each individual.
 * The basis of the rating is the value of the individual minus a
 * minimum value. The way in which the minimum value is determined
 * is governed by the scaling window (w). If w = 0, then the minimum
 * from the first generation is taken for all generations.
 * For 0 < w < 7, the minimum from the last w generations is used.
 * If w = 7, then the minimum possible value is taken. */
void fitness ()  /*int adjust*/
{
      int rc;
      double bestvalue, fmin1, fmintemp, partsum1, rating_scale;
      if (window==7)
            fmin1=0.0;
      else
          if (window == 0)
          {
                if (gencount == 1)
                    fmin1 = 0.0;
                fmintemp = maxpossible;
                for (rc = 0; rc < maxindiv; rc++)
                    if (value[rc] < fmintemp)
                        fmintemp = value[rc];
                if (fmintemp > fmin1)
                    fmin1 = fmintemp;
          }
          else
          {
                if (window > 1)
                    for (rc = 1;rc < window; rc++)
                        fmin[rc] = fmin[rc - 1];
```

```
                fmin[0] = maxpossible;
                for (rc = 0;rc < maxindiv; rc++)
                        if (value[rc] < fmin[0])
                                fmin[0] = value[rc];
                fmin1 = maxpossible;
                for (rc = 0; rc < window; rc++)
                        if (fmin[rc] < fmin1)
                                fmin1 = fmin[rc];
        }
        if (fmin1 == maxpossible)
                fmin1 = 0.0;
        rating_scale = maxpossible / (maxpossible - fmin1);
        bestvalue = 0.0;
        genstrength = 0.0;
        partsum1 = 0.0;
        for (rc = 0;rc < maxindiv; rc++)
        {
                rating[rc] = (value[rc] - fmin1) * rating_scale;
                if (rating[rc] > 0)
                        partsum1 += rating[rc];
                partsum[rc] = partsum1;
                if (value[rc] > bestvalue)
                {
                        bestvalue = value[rc];
                        best = rc;
                }
                genstrength += value[rc];
        }
        genstrength *= gen_scale;
}

/* this procedure is used if the elitist selection strategy is
 * invoked in order to ensure that the fittest member of a
 * generation always survives intact into the next generation */
void elite()
{
    int i = 0, j, nominee;
    int found = 0;
    do
    {
        found = 1;
        for (j = 0; j < words; j++)
        if (source[j][best] != pool[j][i])
                found = 0;
        i += 1;
    } while ((found == 0) && (i < maxindiv));
    if (found == 0)
```

```
        {
            nominee = range_rand(0, maxindiv - 1);
            for (j = 0; j < words; j++)
                pool[j][nominee] = source[j][best];
        }
}


void main()
{
    int i, j, j1, indcounter, n, k;
    unsigned long temp;
    time_t start, finish,t;
    FILE *stream;
    randomize();
    clrscr();
    for(FileNum=0;FileNum<NumOfTrials;FileNum++)
    {
        srand((FileNum+1)*time(NULL));
        itoa(FileNum,FileStep,10);
        strcpy(FileName,FileName1);
        strcat(FileName,FileStep);
        stream = fopen(FileName , "w+");
        start = time(NULL);
        /* initialise constants */
        words = maxbits/32;
        if (maxbits%32 > 0)
        {
            last = maxbits - 32 * words;
            words += 1;
        }
        else
            last = 0;
        /* initialise variables */
        crossmask = 0;
        for (maskcount = 0; maskcount < 32; maskcount++)
        {
            crossmask = (crossmask << 1) | 1;
                *(lower + maskcount) = crossmask;
                *(upper + maskcount) = (long)(-1) ^ crossmask;
        }
        if ((window > 0) && (window < 7))
            for (j = 0; j < window; j++)
                fmin[j] = 0.0;
            gencut1 = (int)(maxindiv * (1.0 - gengap));
        gencut2 = gencut1 + 1;
        gencount = 0;
```

```
        gen_scale = 100.0 / (maxindiv * maxpossible);
        /* generate the population using the random number generator */
        randomize();
        for (j = 0; j < maxindiv; j++)
        {
            for (k = 0; k < words; k++)
            pool[k][j] = (unsigned long)( unit_rand() * two32 - two31);
            if (last>0)
                pool[words-1][j] = pool[words-1][j] & lower[last-1];
        }

        /* optimisation loop */
        while (gencount<maxgen)
        {
            gencount += 1;
            if (gencount > 1)
            {
                for (i=0; i < MAXWORDS; i++)
                {
                    for (j=0; j < maxindiv; j++)
                    {
                        temp = source[i][j];
                        source[i][j] = pool[i][j];
                        pool[i][j] = temp;
                    }
                }
                random_selection();
                seeded_selection();
                crossover();
                mutation();
                if (strategy == 1)
                    elite();
            }
            convert();
            fitness();
            EvalStr = pool[0][best] ;
            fit=evalfunc(EvalStr);
            fprintf(stream,"gencount: %d fit: %f x1= %f x2=%f\n",
            gencount,fit,x[0],x[1]);
        }
        fclose(stream);
    }// for FileNum
} // main
```

Appendix 4

Tabu Search Program

```
/*A standard tabu search algorithm */
#include <stdlib.h>
#include <string.h>
#include <stdio.h>
#include <time.h>
#include <dos.h>
#include <math.h>

#define freqfact 2
#define recfact 5
#define NumIter 2500
#define NumOfVar 2
#define NumOfTrials 10
#define NumOfNeigh 2*NumOfVar
#define Resolution 0.001
#define MIN(x,y) ( (x) < (y) ? (x) : (y) )
#define ABS(x) ( (x) > (0) ? (x): (-(x)) )

int iter,FileNum,itcounter,AllTabu;
int TabuList[NumOfVar],i,j,k,NumEval,BestPos,minRecency,
        minRecIndex,minFreq,minFreqIndex;
long int SuccessChange;
double minimum,TotEvalNum,MinBest,LastBest,
        LastIter,avefreq,TuneCoef;
double recency[NumOfVar],freq[NumOfVar],VarBounds[2*NumOfVar];
double Neigh[NumOfNeigh*NumOfVar],Delta[NumOfVar*NumOfNeigh],
LastResult[NumOfVar];
double x[NumOfVar],y[NumOfVar], Best[NumOfVar],
        result[NumOfNeigh*NumOfVar], EvalNeigh[NumOfVar];
char FileStep[2];
char FileName[12];
char *FileName1="tabu2_";

double evalfunc(double xEval[NumOfVar])
{
```

```
        double evalRes;
        int VarNum;
        evalRes=100.0*pow(pow(xEval[0],2.0)-xEval[1],2.0)+
                pow(1.0-xEval[0],2.0);
        return(evalRes);
}


void initialise()
{
        minimum=0.0;
        TotEvalNum=0.0;
        MinBest=1000.0;
        LastBest=1000.0;
        LastIter=1.0;
        TuneCoef=5.0;
        SuccessChange=0;
        for(i=0;i<NumOfVar;i++)
        {
                recency[i] = 0.0;
                freq[i] = 0.0;
                TabuList[i] = 0;
                VarBounds[2*i ] = -2.048;
                VarBounds[2*i+1]= 2.047;
                x[i] = (double) VarBounds[2*i+1]*
                        ((2*random(1/Resolution)/
                        (1/Resolution) )-1);
                LastResult[i] = x[i];
                result[2*i] = 0.0;
                result[2*i+1] = 0.0;
        };
}


void ProduceNeighbors()
{
        NumEval=0;
        for (i=0;i<NumOfVar;i++)
        {
                for (j=0;j<NumOfNeigh;j++)
                {
                        if (TabuList[i]==0)
                        {
                                do
                                {
                                        Delta[NumOfNeigh*i+j]= MIN(sqrt(2)*
                                        (TuneCoef+Resolution),
```

```
                                sqrt(2)*VarBounds[2*i+1]-
                                VarBounds[2*i])*
                                (random((1/Resolution)) -
                                random((1/Resolution)))/
                                (1/Resolution);
                        }
                        while ( ((x[i]+Delta[NumOfNeigh*i+j] <
                                VarBounds[2*i]) ||
                                (x[i]+Delta[NumOfNeigh*i+j] >
                                VarBounds[2*i+1])));
                        Neigh[NumEval]=x[i]+
                                Delta[NumOfNeigh*i+j];
                        NumEval++;
                }; // if TabuList
        }; // for j = -> NumOfNeigh
    }; // for i=0 -> NumOfVar
    TotEvalNum+=NumEval;
}


void EvaluatingNeighbors()
{
        for(i=0;i<NumOfVar*NumOfNeigh;i++)
        {
                for (j=0;j<NumOfVar;j++)
                        if ((i >= j*NumOfNeigh) &&
                                (i < (j+1)*NumOfNeigh))
                                EvalNeigh[j]=Neigh[i];
                        else
                                EvalNeigh[j]=x[j];
                result[i]=evalfunc(EvalNeigh);
                if (LastBest > result[i])
                        SuccessChange++;
                if (MinBest > result[i])
                {
                        MinBest=result[i];
                        BestPos=i;
                };
        };
}


void UpdateChange()
{
        if (SuccessChange/TotEvalNum > 0.2)
                TuneCoef *= 0.82;
        else
```

```
                    if (SuccessChange/TotEvalNum < 0.2)
                            TuneCoef *= 1.22;
}


void ChooseBestNeighbor()
{
        for (j=0;j<NumOfVar;j++)
                if ((BestPos >= j*NumOfNeigh) &&
                            (BestPos < (j+1)*NumOfNeigh))
                        y[j]=Neigh[BestPos];
                else
                        y[j]=x[j];
                if (LastBest >= MinBest)
                {
                        LastBest=MinBest;
                        LastIter=iter;
                        for (j=0;j<NumOfVar;j++)
                                LastResult[j]=y[j];
                }
                else
                        for (j=0;j<NumOfVar;j++)
                                y[j]=LastResult[j];
}


void UpdateMemory()
{
        for (i=0;i<NumOfVar;i++)
        {
                if (!(x[i] == y[i]))
                {
                        recency[i]=iter;
                        freq[i]++;
                };
        };
        avefreq=0.0;
        for (i=0;i<NumOfVar;i++)
                avefreq+=freq[i];
        avefreq=avefreq/NumOfVar;
        if (iter > 1)
        {
                for (i=0;i<NumOfVar;i++)
                        if (NumOfVar>4)
                        {
                                if (((iter-recency[i])
                                        <= recfact*NumOfVar)
```

```
                                                || (freq[i]
                                                > freqfact*(avefreq+1)))
                        if (((iter-recency[i])
                                        <= recfact*NumOfVar)
                                        || (freq[i]
                                        > freqfact*(avefreq+1)))
                                TabuList[i]=1;
                        else
                                TabuList[i]=0;
                }
                else
                        if (freq[i] > freqfact*(avefreq+1))
                                TabuList[i]=1;
                        else
                                TabuList[i]=0;
        };
}


void Aspiration()
{
        AllTabu=1;
        for (i=0;i<NumOfVar;i++)
        {
                x[i]=y[i];
                if (TabuList[i]==0)
                {
                        AllTabu=0;
                };
        };
        if (AllTabu)
        {
                // Make the least tabu index NotTabu
                minRecency=iter;
                minFreq=100*avefreq;
                minRecIndex=1;
                minFreqIndex=1;
                if ((iter-recency[i]) <= recfact*NumOfVar)
                {
                        if (minRecency > (iter-recency[i]))
                        {
                                minRecency=iter-recency[i];
                                minRecIndex=i;
                        }
                        else
                                if ((freq[i] > freqfact*avefreq)
                                        && (minFreq > freq[i]))
```

```
                                        {
                                        minFreq=freq[i];
                                        minFreqIndex=i;
                                };
                                recency[minRecIndex]=recency[minRecIndex]-1;
                                freq[minFreqIndex]=freq[minFreqIndex]-avefreq;
                                // if it traps at recency
                                TabuList[minRecIndex]=0;
                                TabuList[minFreqIndex]=0;
                        }
                } // if TabuList ==0
}


void main()
{
        FILE *stream;
        time_t t;
        randomize();
        clrscr();
        for(FileNum=0;FileNum<NumOfTrials;FileNum++)
        {
                srand((FileNum+1)*time(NULL));
                itoa(FileNum,FileStep,10);
                strcpy(FileName,FileName1);
                strcat(FileName,FileStep);
                stream = fopen(FileName , "w+");
                initialise();
                itcounter=0;
                for(iter=0;iter<NumIter;iter++)
                {
                        ProduceNeighbors();
                        EvaluatingNeighbors();
                        UpdateChange();
                        ChooseBestNeighbor();
                        UpdateMemory();
                        Aspiration();
                        itcounter+=itcounter;
                        if (itcounter==10)
                        {
                                SuccessChange=0;
                                TotEvalNum=0;
                                itcounter=0;
                        }
                        printf("\n %f - %d - %f - %f",LastBest,iter,x[0],x[1]);
                        fprintf(stream,"\n %f %f %f ",LastBest,x[0],x[1]);
                }; //for iter=0 -> NumIter
```

```
            fclose(stream);
        }// for FileNum
} // main
```

Appendix 5

Simulated Annealing Program

```
/*A standard simulated annealing algorithm*/
#include <stdlib.h>
#include <string.h>
#include <stdio.h>
#include <time.h>
#include <dos.h>
#include <math.h>

#define NumIter 500
#define NumOfVar 2
#define NumOfTrials 10
#define NumOfNeigh 2*NumOfVar
#define Resolution 0.001
#define StartValueOfT 100.0
#define nRep 10
#define MIN(x,y) ( (x) < (y) ? (x) : (y) )
#define ABS(x) ( (x) > (0) ? (x): (-(x)) )

unsigned int iter,FileNum;
unsigned int i,j,k,NumEval,BestPos;
unsigned int SuccessChange;
double minimum,TotEvalNum,MinBest,LastBest,LastIter,TuneCoef,T;
double VarBounds[2*NumOfVar];
double Neigh[NumOfNeigh*NumOfVar],Delta[NumOfVar*NumOfNeigh],
    LastResult[NumOfVar];
double x[NumOfVar],y[NumOfVar],Best[NumOfVar],
    result[NumOfNeigh*NumOfVar],EvalNeigh[NumOfVar];
double Difference;
char FileStep[2];
char FileName[12];
char *FileName1="sim2_";

double evalfunc(double xEval[NumOfVar])
{
    double evalRes;
```

```
    evalRes=100.0*pow(pow(xEval[0],2.0)-xEval[1],2.0)+
        pow(1.0-xEval[0],2.0);
    return(evalRes);
}


void initialise()
{
    minimum=0.0;
    TotEvalNum=0.0;
    LastBest=1000.0;
    LastIter=1.0;
    TuneCoef=5.0;
    T=StartValueOfT;
    SuccessChange=0;
    for(i=0;i<NumOfVar;i++)
    {
        VarBounds[2*i ] = -2.048;
        VarBounds[2*i+1]= 2.047;
        x[i] = (double) VarBounds[2*i+1]*((2*random(1/Resolution)/
            (1/Resolution) )-1);
        LastResult[i] = x[i];
        result[2*i] = 0.0;
        result[2*i+1] = 0.0;
    }
}


void ProduceNeighbors()
{
    NumEval=0;
    for (i=0;i<NumOfVar;i++)
    {
        for (j=0;j<NumOfNeigh;j++)
        {
            do
            {
                Delta[NumOfNeigh*i+j]=
                MIN(sqrt(2)*(TuneCoef+Resolution),
                sqrt(2)*VarBounds[2*i+1]-VarBounds[2*i])*
                (random((1/Resolution)) - random((1/Resolution)))/
                (1/Resolution);
            } while ( ((x[i]+Delta[NumOfNeigh*i+j] < VarBounds[2*i])
                || (x[i]+Delta[NumOfNeigh*i+j] > VarBounds[2*i+1])));
            Neigh[NumOfNeigh*i+j]=x[i]+Delta[NumOfNeigh*i+j];
            NumEval++;
        }; // for j = -> NumOfNeigh
```

```
    }; // for i=0 -> NumOfVar
    TotEvalNum+=NumEval;
}


void EvaluatingNeighbors()
{
    MinBest=100000.0;
    for(i=0;i<NumOfVar*NumOfNeigh;i++)
    {
        for (j=0;j<NumOfVar;j++)
        if ((i >= j*NumOfNeigh) && (i < (j+1)*NumOfNeigh))
            EvalNeigh[j]=Neigh[i];
        else
            EvalNeigh[j]=x[j];
        result[i]=evalfunc(EvalNeigh);
        if (LastBest > result[i])
            SuccessChange++;
        if (MinBest > result[i])
        {
            MinBest=result[i];
            BestPos=i;
        }
    };
}


void UpdateChange()
{
    if (SuccessChange/TotEvalNum > 0.2)
        TuneCoef *= 0.82;
    else
    if (SuccessChange/TotEvalNum < 0.2)
        TuneCoef *= 1.22;
    TuneCoef=MIN(TuneCoef,2.047); /* VarBounds[1] */
}


void ChooseBestNeighbor()
{
    for (j=0;j<NumOfVar;j++)
      if ((BestPos >= j*NumOfNeigh) && (BestPos < (j+1)*NumOfNeigh))
            y[j]=Neigh[BestPos];
        else
            y[j]=x[j];
        if (LastBest >= MinBest)
        {
```

```
                    LastBest=MinBest;
                    LastIter=iter;
                    for (j=0;j<NumOfVar;j++)
                            LastResult[j]=y[j];
            }
            else
            {
                    Difference=evalfunc(x)-MinBest;
                    if ((Difference < 0) ||
                            ((Difference > 0) && (random(1) <
                            exp((-1.0)*(Difference/T)))))
                    {
                    for (j=0;j<NumOfVar;j++)
                            if ((BestPos >= j*NumOfNeigh) &&
                                        (BestPos < (j+1)*NumOfNeigh))
                                    y[j]=Neigh[BestPos];
                            else
                                    y[j]=x[j];
                    }
                    else
                            for (j=0;j<NumOfVar;j++)
                                    y[j]=x[j];
            };
            for (j=0;j<NumOfVar;j++)
                    x[j]=y[j];
}


void main()
{
    FILE *stream;
    time_t t;
    int InnerCounter;
    randomize();
    clrscr();
    for(FileNum=0;FileNum<NumOfTrials;FileNum++)
    {
        srand((FileNum+1)*time(NULL));
        itoa(FileNum,FileStep,10);
        strcpy(FileName,FileName1);
        strcat(FileName,FileStep);
        stream = fopen(FileName , "w+");
        initialise();
        for(iter=0; iter < NumIter; iter++)
        {
            for(InnerCounter=0;InnerCounter<nRep;InnerCounter++)
            {
```

```
                      ProduceNeighbors();
                      EvaluatingNeighbors();
                      UpdateChange();
                      ChooseBestNeighbor();
                 }
                 T = 0.1*T;
                 fprintf(stream,"%f %f %f %f \n",LastBest,T,x[0],x[1]);
            }; //for iter=0 -> NumIter
            fclose(stream);
      }// for FileNum
} // main
```

Appendix 6

Neural Network Programs

```c
#include <math.h>
#include <stdio.h>
#include <stdlib.h>
#include <conio.h>

#define N 3       // the number of neurons

void main()
{
 float E, x[N];
 float delt;
 float factor;
 float a = 5.12;
 float y;
 int   k; // neuron
 int   coverage;
 int   flag[N];

//---- set initial value
 printf("ASSIGN initial value {0, 1} TO EVERY NEURON\n");
 for(k = 0; k < N; k ++){
  printf("neuron_%d = ", k);
  scanf("%f", &x[k]);
  x[k] = x[k]/a;
 }

//---- initilisation
 coverage = 0;
 factor = 0.1;
 y = 0;
 for(k = 0; k < N; k ++)
  flag[k] = 0;

//---- random upgrade Hopfield outputs & energy function
 do{
```

```
// random choose one neuron
while(!coverage){
 k = random(N);
 if(!flag[k])
  coverage = 1;
}
flag[k]  = 1;  // set flag k
coverage = 0;  // release coverage
//update
if(k == 0)
 delt = factor*(2*x[0]+x[1]+x[2]);
if(k == 1)
 delt = factor*(2*x[1]+x[2]+x[0]);
if(k == 2)
 delt = factor*(2*x[2]+x[0]+x[1]);

x[k] = x[k] - delt;
if(x[k] <= -1)
 x[k] = -1;
if(x[k] >= 1)
 x[k] = 1;

if(N == flag[0]+flag[1]+flag[2]){
 for(k = 0; k < N; k ++){          // release all flags
  flag[k] = 0;
  printf("x[%d]=%f,", k, x[k]);
 }
 E = a*a*(0.5*(x[0]*x[0] + x[1]*x[1] +x[2]*x[2])
  -(x[0]*x[1] + x[1]*x[2] + x[0]*x[2]));
 printf("E=%f\n", E);
 }
}
while(!kbhit());

//---- export
x[0] = a*x[0];
x[1] = a*x[1];
x[2] = a*x[2];

y = x[0]*x[0] + x[1]*x[1] + x[2]*x[2];  // compute cost function
printf("x[0]=%f\tx[1]=%f\tx[2]=%f\n", x[0],x[1],x[2]);
printf("y=%f\n", y);

return;
}
```

```c
#include <math.h>
#include <stdio.h>
#include <stdlib.h>
#include <conio.h>

#define N 2      // the number of neurons
#define M 2     // the number of nets

void main()
{
 float E, x[M+N];
 float delt;
 float factor[M+N];
 float a = 2.048;
 float a2, a3, a4, y;
 int   k; // neuron
 int   coverage;
 int   flag[N+M];

//---- set initial value
printf("ASSIGN initial value TO EVERY NEURON\n");
for(k = 0; k < N; k ++){
 printf("neuron_%d = ", k);
 scanf("%f", &x[k]);
}

//---- normilisation
 a2 = a*a;
 a3 = a*a2;
 a4 = a*a3;
 x[0] = (x[0]/a)*(x[0]/a);
 x[1] = x[1]/a;
 x[2] = 1;
 x[3] = x[0]/a;

//---- initilisation
 coverage  = 0;
 factor[0] = 0.01/(100*a2);
 factor[1] = 0.01/(100*a);
 factor[2] = 0;
 factor[3] = 0.01/a;
 y = 0;
 for(k = 0; k < N+M; k ++)
  flag[k] = 0;

//---- random upgrade Hopfield outputs & energy function
do{
```

```
// random choose one neuron
while(!coverage){
 k = random(N);
 if(!flag[k])
  coverage = 1;
}
flag[k]  = 1;  // set flag k
coverage = 0;   // release coverage
//update
if(k == 0)
 delt = factor[k]*(2*a2*x[0]-a*x[1]-1)*100*a2;
if(k == 1)
 delt = factor[k]*(2*a*x[1]-a2*x[0]-1)*100*a;
if(k == 2)
 delt = 0;
if(k == 3){
 x[3] = sqrt(x[0]);
 delt = factor[k]*(2*a*x[3]-x[2]-1)*a;
 x[0] = x[3]*x[3];
}

x[k] = x[k] - delt;
if(x[k] <= -1)
 x[k] = -1;
if(x[k] >= 1)
 x[k] = 1;

if(N == flag[0]+flag[1]+flag[2]+flag[3]){
 for(k = 0; k < N+M; k ++){          // release all flags
  flag[k] = 0;
  printf("x[%d]=%f,", k, x[k]);
 }
 E = 100*(a4*x[0]*x[0] - a3*x[0]*x[1] + a2*x[1]*x[1] -a2*x[0] -a*x[1])
   +(x[2]*x[2] - a*x[2]*x[3] +a2*x[3]*x[3] -x[2] -a*x[3]);
 printf("E=%f\n", E);
}
}
while(!kbhit());

//---- export
x[0] = a2*x[0]; //x0 square
x[1] = a *x[1]; //x1
x[2] = 1;      //x2
x[3] = a *x[3]; //x3

y = 100*(x[0]-x[1])*(x[0]-x[1]) + (1-x[3])*(1-x[3]);  // compute cost function
printf("x0=%f\tx1=%f\n", sqrt(x[0]), x[1]);
```

```
printf("y=%f\n", y);

return;
}
```

```
#include <math.h>
#include <stdio.h>
#include <stdlib.h>
#include <conio.h>

#define N 5      // the number of neurons

void main()
{
 float E;
 float x[N];
 float delt;
 float factor;
 float a = 5.12;
 float y;
 int  k; // neuron
 int  coverage;
 int  flag[N];

//---- set initial value
printf("ASSIGN initial value TO EVERY NEURON\n");
for(k = 0; k < N; k ++){
 printf("neuron_%d = ", k);
 scanf("%f", &x[k]);
 x[k] = x[k]/a;    // normilisation
}

//---- initilisation
coverage = 0;
factor  = 1/a;
y = 0;
for(k = 0; k < N; k ++)
 flag[k] = 0;

//---- random upgrade Hopfield outputs & energy function
do{
 // random choose one neuron
 while(!coverage){
  k = random(N);
  if(!flag[k])
   coverage = 1;
 }
 flag[k]  = 1;  // set flag k
 coverage = 0;  // release coverage
 //update
 if(k == 0)
  delt = factor*(1-x[0]*x[0])*(1+(x[1]+x[2]+x[3]+x[4])/4)*a;
```

```
if(k == 1)
 delt = factor*(1-x[1]*x[1])*(1+(x[0]+x[2]+x[3]+x[4])/4)*a;
if(k == 2)
 delt = factor*(1-x[2]*x[2])*(1+(x[0]+x[1]+x[3]+x[4])/4)*a;
if(k == 3)
 delt = factor*(1-x[3]*x[3])*(1+(x[0]+x[1]+x[2]+x[4])/4)*a;
if(k == 4)
 delt = factor*(1-x[4]*x[4])*(1+(x[0]+x[1]+x[2]+x[3])/4)*a;

x[k] = x[k] - delt;

if(N == flag[0]+flag[1]+flag[2]+flag[3]+flag[4]){
 for(k = 0; k < N; k ++){          // release all flags
  flag[k] = 0;
  printf("x[%d]=%f,", k, x[k]);
 }
 E = (x[0] + x[1] + x[2] + x[3] + x[4])*a
  +(x[0]*x[1] + x[0]*x[2] + x[0]*x[3] + x[0]*x[4]
      + x[1]*x[2] + x[1]*x[3] + x[1]*x[4]
          + x[2]*x[3] + x[2]*x[4]
              + x[3]*x[4])/4;
 printf("E=%f\n", E);
 }
}
while(!kbhit());

//---- export
for(k = 0; k < N; k ++)
 x[k] = a*x[k];

y = x[0]+x[1]+x[2]+x[3]+x[4];   // compute cost function
printf("x0=%f\nx1=%f\nx2=%f\nx3=%f\nx4=%f\n",
   x[0], x[1], x[2], x[3], x[4]);
printf("y=%f\n", y);

return;
}
```

```c
#include <math.h>
#include <stdio.h>
#include <stdlib.h>
#include <conio.h>

#define N 6       // the number of neurons

void main()
{
 float E, x[N];
 float delt;
 float factor;
 float a = 65.536;
 float a11, a12;
 float y;
 int   k; // neuron
 int   coverage;
 int   flag[N];

//---- set initial value
 printf("ASSIGN initial value TO EVERY NEURON\n");
 for(k = 0; k < N; k ++){
  if(0 == fmod(k, 3)){
   printf("neuron_%d = ", k);
   scanf("%f", &x[k]);
   x[k] = x[k]/a;
  }
 }
 x[1] = pow(x[0], 2);
 x[2] = pow(x[0], 3);
//---- initilisation
 y  = 0;
 a11 = 32;
 a12 = 32;
 factor = 0.01/(a*a*a*a*a);
 coverage = 0;
 for(k = 0; k < N; k ++)
  flag[k] = 0;

//---- random upgrade Hopfield outputs & energy function
 do{
 // random choose one neuron
  while(!coverage){
  k = random(N);
  if(!flag[k])
   coverage = 1;
  }
```

```
flag[k] = 1;  // set flag k
coverage = 0;  // release coverage
//update
if(k == 0)
 delt = factor*10*a*a11*a11*(a*a11*a11*x[0]+a*a*a11*x[1]+
         a*a*a*x[2]+0.6*a11*a11*a11);
if(k == 1)
 delt = factor*10*a*a*a11*(a*a11*a11*x[0]+a*a*a11*x[1]+
         0.6*a*a*a*x[2]+a11*a11*a11);
if(k == 2)
 delt = factor*10*a*a*a*(a*a11*a11*x[0]+0.6*a*a*a11*x[1]+
         0.2*a*a*a*x[2]+a11*a11*a11);
/*if(k == 3)
 delt = factor*10*a*a12*a12*(a*a12*a12*x[3]+a*a*a12*x[4]+
         a*a*a*x[5]+0.6*a12*a12*a12);
if(k == 4){
 x[4] = x[3]*x[3];
 delt = factor*10*a*a*a12*(a*a12*a12*x[3]+a*a*a12*x[4]+
         0.6*a*a*a*x[5]+a12*a12*a12);
 x[3] = sqrt(x[4]);
}
if(k == 5){
 x[5] = x[3]*x[3]*x[3];
 delt = factor*10*a*a*a*(a*a12*a12*x[3]+0.6*a*a*a12*x[4]+
         0.2*a*a*a*x[5]+a12*a12*a12);
 x[3] = pow(x[5], 1/3);
 x[4] = pow(x[5], 2/3);
}
   */
x[k] = x[k] - delt;
if(x[k] <= -1)
 x[k] = -1;
if(x[k] >= 1)
 x[k] = 1;

if(N == flag[0]+flag[1]+flag[2]+flag[3]+flag[4]+flag[5]){
 for(k = 0; k < N; k ++){              // release all flags
  flag[k] = 0;
  if(0 == fmod(k, 3))
   printf("x[%d]=%f,", k, x[k]);
 }
 E = 0.001*(5*a*a*a11*a11*a11*a11*x[0]*x[0] +
    5*a*a*a*a*a11*a11*x[1]*x[1] +
    a*a*a*a*a*x[2]*x[2] +
    10*a*a*a*a11*a11*a11*x[0]*x[1] +
    10*a*a*a*a*a11*a11*x[0]*x[2] +
    6*a*a*a*a*a*a11*x[1]*x[2] +
```

```
      6*a*a11*a11*a11*a11*a11*x[0] +
      10*a*a*a11*a11*a11*a11*x[1] +
      10*a*a*a*a11*a11*a11*x[2] +
      5*a*a*a12*a12*a12*a12*x[3]*x[3] +
      5*a*a*a*a*a12*a12*x[4]*x[4] +
      a*a*a*a*a*a*x[5]*x[5] +
      10*a*a*a*a12*a12*a12*x[3]*x[4] +
      10*a*a*a*a*a12*a12*x[3]*x[5] +
      6*a*a*a*a*a*a12*x[4]*x[5] +
      6*a*a12*a12*a12*a12*a12*x[3] +
      10*a*a*a12*a12*a12*a12*x[4] +
      10*a*a*a*a12*a12*a12*x[5]);
   printf("E=%f\n\n", E);
  }
 }
 while(!kbhit());

 //---- export
 x[0] = a*x[0];
 x[3] = a*x[3];

 y = 1 + pow((x[0]-a11), 6) + pow((x[3]-a12), 6);   // compute cost function
 printf("x[0]=%f\tx[1]=%f\n", x[0],x[3]);
 printf("y=%f\n", y);

 return;
 }
```

Author Index

Subject Index